Y0-DJO-020

Volume 5

Essays in Physics

Volume 5
1973

Essays in Physics

Edited by

G.K.T. Conn and **G.N. Fowler**
Department of Physics
The University of Exeter
Exeter, U.K.

Academic Press London and New York

A Subsidiary of Harcourt Brace Jovanovich, Publishers

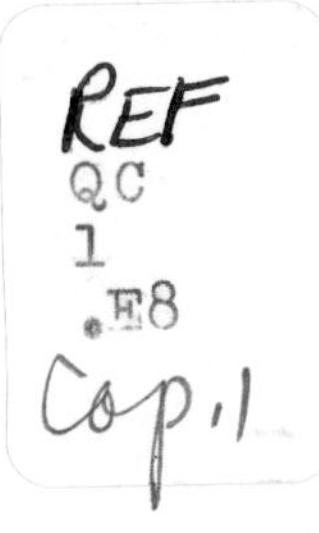

ACADEMIC PRESS INC. (LONDON) LTD.
24/28 Oval Road,
London NW1

United States Edition published by
ACADEMIC PRESS INC.
111 Fifth Avenue
New York, New York 10003

Library of Congress Catalog Card Number: 74-117120
ISBN: 0-12-184805-1

Printed in Northern Ireland at The Universities Press, Belfast

List of Contributors

ROBIN BRETT, Lyndon B. Johnson Space Center, NASA, Houston, Texas, U.S.A. (p. 1)

G. H. A. COLE, Physics Department, University of Hull, Hull, England. (p. 37)

D. SETTE, Istituto di Fisica, Facolta di Ingegneria, Universita di Roma, Roma, Italy, and Gruppo Nazionale di Struttura della Materia del Consiglio Nazionale Delle Ricerche, Roma, Italy. (p. 95)

JOHN STRONG, Astronomy Research Facility, University of Massachusetts, Amherst, Massachusetts, U.S.A. (p. 197)

Preface

In 1873 van der Waals published his equation of state. It was a fundamental step in understanding the behaviour of real gases and lead to the concept of "critical states." Only in the last year or two has further progress been made of comparable importance, the development of scaling applied to critical phenomena. The review of Critical Phenomena by Professor Sette is therefore very appropriate in this centenary year.

Another step forward of a very different kind was that by Armstrong and Aldrin when they landed on the moon in 1969. This is described by Dr. Brett in this issue as "the most intensive multi-disciplinary, multi-national co-operative research program ever undertaken." We must expect further evidence from deep space probes in the not too distant future hence Professor Cole's essay on the Physics of Planetary Interiors. For those whose feet are firmly fixed on this planet the important conclusion "that the concept of instrument-limited resolving power is thus obsolete and the only limitation left is the true one—the energy-limited resolving power" emerging from Professor Strong's essay on Multiplex Spectrometry will surprise many. How long will it be before the teaching texts report this?

G. K. T. Conn
G. N. Fowler

Contents

Lunar Science

ROBIN BRETT

Lyndon B. Johnson Space Center, NASA, Houston, Texas U.S.A.

I. Introduction

The moon, which for centuries has been an object of interest to poets, fishermen, astronomers, and lovers, remained inviolate from man until 1969 when astronauts Neil Armstrong and Edwin Aldrin set foot on its surface. Since then there have been five additional Apollo landings and two Soviet unmanned sample returns (Fig. 1). The lunar samples and geophysical experiments of the Apollo program have resulted in the most intensive multi-disciplinary, multi-national cooperative research program ever undertaken in any field of science, with the result that the published literature of lunar science now exceeds 10,000 printed pages.

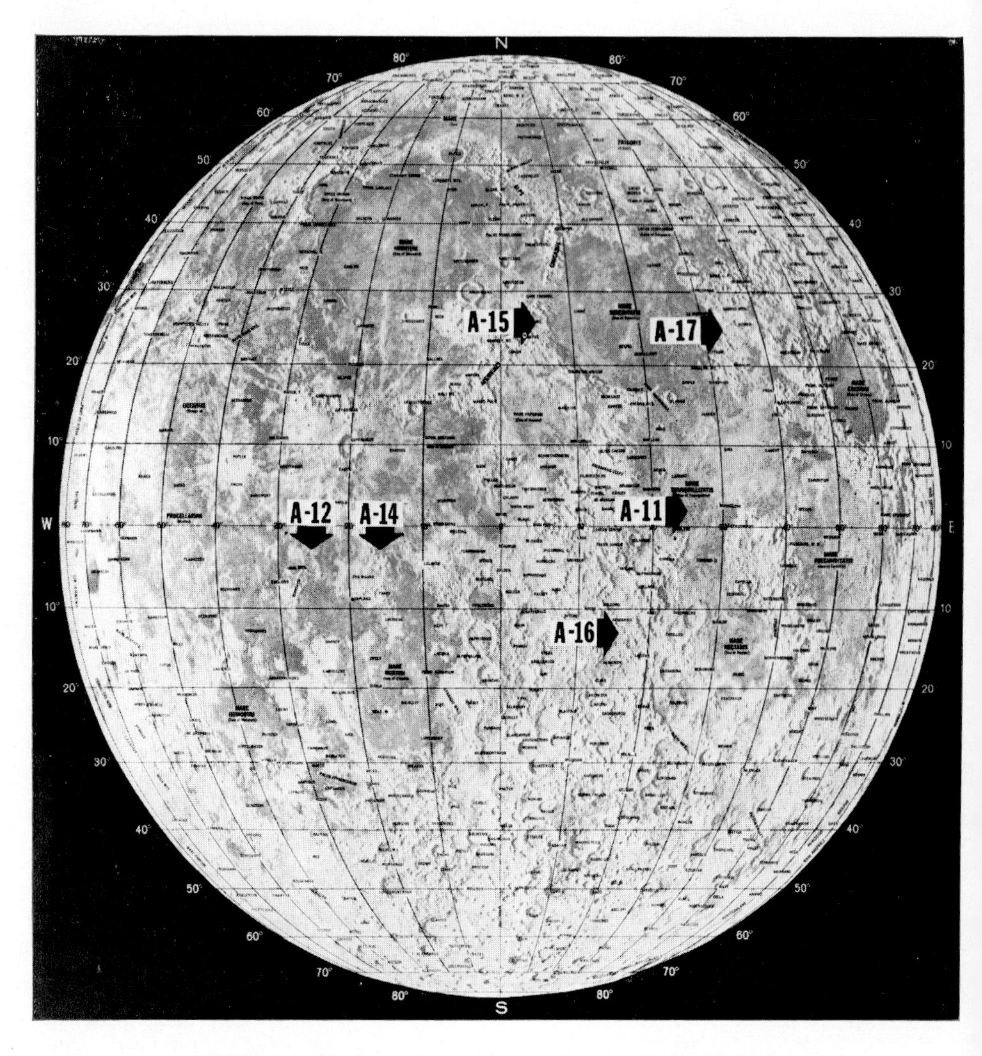

FIGURE 1. The lunar front side, showing Apollo landing sites.

The main reasons for studying the moon are threefold: (1) to understand the evolution of a planet other than the earth, (2) to obtain the bulk composition and structure of another body in the solar system and thus to define principles of chemical differentiation within the solar nebula, and (3) to examine in detail the history of the first one and a half billion years of lunar history with the view to understand better the same period in the history of our own planet. The record of that early period on earth has been erased by the earth's much more active subsequent history.

The origin of the moon is still not understood nor is its evolution known in great detail. However, the advance in understanding is prodigious since July, 1969. For example, prior to that date some scientists believed that the lunar maria were water-lain sediments and others concerned NASA safety engineers by proposing that lunar soil would catch fire when brought into the oxygen-rich atmosphere of the Lunar Module.

It is the purpose of the present review to summarize the more important lunar findings and the known restraints on theories of lunar evolution. Much interesting work is by necessity omitted owing to limitations of space. Similarly, I have not attempted to discuss the extremely important conclusions regarding the solar wind and cosmic rays which have been by-products of lunar research, but have confined this review to the evolution of the moon itself. The reader is referred to Hinners (1971), Urey and MacDonald (1971), and Lowman (1972) for other recent reviews on the subject. Each of these reviews has its own particular emphasis; Hinners' work is perhaps the most general of the three.

II. Lunar Geophysics

A. FIGURE OF THE MOON

The bulk density of the moon is 3.35 gr/cm^3 (Kaula, 1971a) and all geochemical models of the lunar interior must meet this restraint. The coefficient of moment of inertia of the moon is 0.402 ± 0.002 (Kaula, 1970). This compares with 0.4 for a sphere of uniform density and 0.67 for a spherical shell. It therefore appears that the density of the moon remains relatively constant with depth and that if the moon has a molten iron-rich core, it must occupy less than about 20 percent of the lunar radius (Runcorn *et al.*, 1970). The moments of inertia and irregular shape of the moon have also led some workers, e.g. McDonald and Urey (1971) and Kopal (1972), to postulate that considerable stresses must be present in the interior of the moon, thus indicating a rigid and therefore cool interior.

A laser altimeter on the Apollo 15 and 16 missions greatly increased the amount of data available on lunar surface topography. Data analyses by Wollenhaupt and Sjogren (1972) demonstrate conclusively that there is evidence for a displacement between the center of figure and center of mass of the moon along the earth-moon line of between 1.5 and 3.8 km.

B. MASCONS

Areas of the moon with positive gravity anomalies (mascons) have been known for some time (Muller and Sjogren, 1968). They are confined almost exclusively to circular mare basins greater than 200 km in diameter (Fig. 2)

FIGURE 2. Mascons on the lunar front side (after Muller and Sjogren, 1968).

and are commonly surrounded by a small negative anomaly. Observation of these anomalies was greatly improved by Apollo missions that carried an S-band transponder thereby allowing precision Doppler tracking of the spacecraft's accelerations and decelerations due to the changes in gravity. Data from the Apollo 15 mission strongly suggest that the mascons are caused by features near the surface that have an excess mass distribution per unit area of approximately 800 kg/cm² (Sjogren *et al.*, 1972).

The origin of mascons has been a controversial subject since they were discovered. Wise and Yates (1970) and Wood (1970) argue that a plug of dense material flowed plastically upward to achieve isostatic compensation of the mare basins and that the later basalts which fill these basins formed the mascons. Kaula (1969) suggests that thermal contraction of the moon's upper surface generated excess pressure to drive dense plugs of mantle material beyond the isostatic level, thus forming the mascons. Urey (1968) proposes that the mascons are remnants of the colliding objects which produced the circular basins. H. Masursky, in numerous lectures, proposes that mascons represent compensation greater than isostatic caused by the formation of a central uplift of high density mantle material as a consequence of the impact process. Dense lunar mantle would not be involved in the central uplift in smaller craters, so such craters should not contain mascons. The negative anomalies surrounding mascons are explained by analogy with the peripheral synclines that characteristically surround terrestrial central uplifts. A similar phenomenon on a large scale on the moon would cause depression of dense mantle material, thus producing a negative anomaly.

The present author independently reached the same conclusion as Masursky, but recently modified it because Mare Orientale, a large impact basin with little mare fill, contains a mascon only in the vicinity of the mare fill (P. Muller, oral communication). Muller also states that mascons are flat, plate-like bodies near the surface. These facts suggest that mascons are at least partially due to mare fill from considerable depth. This theory therefore embraces portions of the theories of Masursky, Wood (1970) and Wise and Yates (1970).

Urey has repeatedly pointed out that mascons are significant to our understanding of lunar thermal history as they indicate that the outer portion of the moon, the so-called lithosphere, was sufficiently rigid from the time they were formed (c. 3.5 b.y. ago), and therefore cool, to support them. A lithosphere of at least 150 km thickness is required (Kaula, 1969).

C. THE LUNAR THERMAL REGIME

An understanding of the moon's thermal history is important, not only in interpreting its igneous activity and hence chemical differentiation, but also in interpreting its figure, its internal stresses and their compensation, and ultimately the rate of accretion of the moon.

Two lunar surface experiments have provided data on interior temperatures and heat flow which contribute greatly to an understanding of the thermal history. The heat flow experiment carried out on the Apollo 15 mission (Langseth *et al.*, 1972) gave the heat flow from depth as 3.3×10^{-6} w/cm^2, a value approximately half that of average heat flow from the earth. This surprisingly high result has led Langseth *et al.* (1972) to conclude that if the heat flow data are representative, then the moon must be more radioactive than either the earth or chondrites. This conclusion need not be valid if the distribution of radioactive elements in the moon is not uniform with depth. The presence of a rock type (KREEP) rich in K, U, and Th and its early formation age (4.4 b.y.; Schonfeld, 1972) suggests that the moon has been heterogeneous for a long time.

The heat flow results are higher than would be predicted by recent thermal models, which led McConnell and Gast (1972) to suggest that either (1) lunar radioactivity is higher than that assumed by most models, or (2) a highly effective heat transfer, perhaps due to solid state convection, is causing the interior to cool rapidly at present.

The lunar surface magnetometer experiments (e.g. Dyal and Parkin, 1972) allow estimation of a temperature profile for assumed compositions in the interior of the moon. When the moon is located in the free streaming solar wind, eddy currents are induced within the moon due to transient phenomena in the solar magnetic field. A surface magnetometer can be used during lunar night to determine the time-dependant decay characteristics of the poloidal field. These decay characteristics are a function of the electrical conductivity in the lunar interior which is itself largely a function of composition and temperature. Therefore, if a given composition is assumed for the lunar interior, then a temperature profile can be obtained (Dyal and Parkin, 1972).

The data obtained are remarkably close to those predicted by the classical theory of a conducting sphere in a magnetic field. The simplest model to explain the magnetic field step transient measurements consists of a spherically symmetrical 3 layered moon in which the non-conducting outer crust has a radius of $0.03R_{\mathrm{moon}}$ (c. 50 km), and the intermediate layer has a thickness of about $0.07R_{\mathrm{moon}}$ (c. 640 km). Dyal and Parkin (1972) and several other workers have calculated temperature profiles based on this model. Unfortunately at present data on the composition of the lunar interior and electrical properties of possible compositions are so scant and conflicting that estimated temperatures have a low degree of confidence.

Thermal models are only as good as the data that are used, and they depend on a large number of variables which are not yet well understood. Among these are the thermal energy produced from the moon's accretion, which is dependant on the duration of the accretion process and the temperature of accreting material, the abundance of primordial isotopes with short-lived radioactivity like ^{26}Al in the young moon, the abundance and distribution within the moon of the long-lived radioisotopes of K, U, and Th, the thermal conductivity and specific heat profile of the moon, the efficiency of conversion of tidal stresses to thermal energy within the moon, cooling due to magma transport, and the thermal energy released by core formation, if, in fact, the moon has a metallic core. Another process, that can dominate thermal models if its existence is proven, is that of solid state convection in the lunar mantle. Tozer (1972) suggests that convection would fairly rapidly bring the interior of the moon to a steady state temperature between 600 and 1000°C.

Thermal models must be compatible with the evolution of the moon, our present understanding of which indicates widespread melting in at least the outer portion of the moon early in its history, and again somewhat later, from about 4 to about 3 b.y. ago when partial melting in the interior yielded lavas that flooded the pre-existing mare basins. After the formation of the maria little external igneous activity occurred, suggesting that at least the outer few hundred kilometers of the moon are solid and relatively cool. The presence of mascons and the non-equilibrium figure of the moon also suggest that at least the outer portion of the moon has been rigid since their formation. The thermal models must also provide for a molten metallic core (e.g. Runcorn *et al.*, 1970) if its existence is proven. Papanastassiou and Wasserburg (1971) discuss the restraints and their implications in detail.

In spite of the inherent problems in reaching satisfactory thermal models for a body like the moon, a number of investigators (e.g. Urey, 1952; MacDonald, 1959; Levin, 1962; Anderson and Phinney, 1967; Fricker *et al.*, 1967; McConnell *et al.*, 1967; Hanks and Anderson, 1969; Wood, 1972; McConnell and Gast, 1972; Toksöz *et al.*, 1972a) have undertaken this task. Most recent models start with a moon cool in the deep interior and at or near melting temperatures near the surface during its first couple of hundred million years of existence. As the effect of long-lived radioactive heating becomes greater than the initial (accretional) heat, temperatures fall near the surface thus causing a cool rigid shell to form and temperatures rise at depths of two to several hundred kilometers, thus intersecting the solidus (the beginning of melting) and causing the generation of basaltic lavas that erupt to fill the maria. The maximum temperature migrates deeper into the moon with time and the center of the moon is well above 1000°C at present according to most recent thermal models (Fig. 3). However, none of these models treats cooling by solid state convection.

Most temperatures calculated from the magnetic measurements are consistent with the theory of convective cooling developed by Tozer (1972), and are also compatible with a moon cool enough to support the mascons and postulated internal stresses (e.g. Kaula, 1969). These temperatures are considerably lower than those predicted by recent thermal models (e.g. McConnell and Gast, 1972; Toksöz *et al.*, 1972a). McConnell and Gast see no solution to the discrepancy unless the original uranium concentration in the interior of the moon was considerably less than is assumed, an assumption which appears to conflict with the heat flow results.

[Added in proof: The heat flow result from the Apollo 17 mission is very similar to that of Apollo 15 (M. Langseth, oral communication, 1973) thus

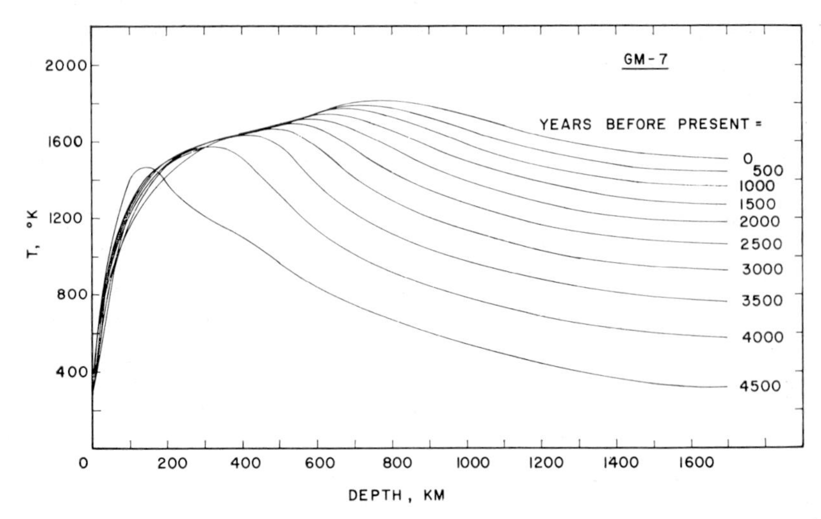

FIGURE 3. A model of lunar thermal history, courtesy of R. J. McConnell. This model is one of several constructed by McConnell and Gast using the method of McConnell and Gast (1972). Time is listed in millions of years.

suggesting that the Apollo 15 result is not anomolous. Also a large meteoroid impact in July, 1972 allowed the seismic network to detect attenuated P waves below 1000 km depth. This attenuation suggests that the moon is at least partially molten below that depth (G. Latham, oral communication, 1972) and that temperatures must be at least as great as 1500°C. It therefore would appear that abundances of radioactive elements assumed in the thermal models are reasonable and that the moon has a hot interior today.]

D. SEISMIC STUDIES

The third lunar seismometer, deployed as part of the passive seismic lunar net by the Apollo 15 mission, established a triangular network and increased the value of the passive seismic experiment considerably because it allowed determination of the source of seismic events (e.g. Latham *et al.*, 1972) and, more important, depth-velocity profiles for the lunar interior.

Nearly 80% of moonquakes apparently derive from a single focus (Latham *et al.*, 1972), located at a depth of about 800 km below an epicenter at Lat. 21°S and Long. 28°W. This depth is considerably greater than those of all earthquake foci and indicates that the lunar mantle at 800 km depth is strong enough to support considerable shear stress. This finding is compatible with a relatively cool lunar interior, in agreement with the magnetic results.

Most moonquakes occur in monthly cycles, a fact that suggests that they are triggered by tidal effects (Latham *et al.*, 1972).

Swarms of small moonquakes have signatures different from those of moonquakes of the monthly cycles. The signals of the lunar swarms are similar to those of terrestrial swarms that occur in volcanic areas before, during, and after eruptions (Latham *et al.*, 1972). It seems unlikely that the moonquakes are associated with near-surface igneous activity since such a conclusion is not compatible with our understanding of the thermal regime in the upper portion of the moon.

Moonquakes are generally very small in magnitude, the largest ever detected registering only 2 to 3 on the Richter scale (Latham *et al.*, 1972). Calculations based on presently available data indicate that the total energy released by moonquakes each year is approximately 10^{11} to 10^{15} ergs, compared with an estimated 5×10^{24} ergs/yr for the earth (Latham *et al.*, 1972). This result suggests to Latham *et al.* that convection currents which might cause significant tectonic activity are probably absent in the lunar interior at present. The possibility remains open that slow convection currents may occur at great depth, below an outer portion of the moon that is cool and rigid.

The seismic data obtained from artificial impacts, have enabled Toksöz *et al.* (1972b) to interpret the structure of the interior of the moon to a depth of about 100 km in the Fra Mauro region of Oceanus Procellarum. They postulate a basaltic outer crust extending to a depth of about 25 km. Below this, sound velocities are nearly constant (6.8 km/sec) to a depth of some 65 km matching the range of velocities of terrestrial gabbros and anorthositic gabbros. The discontinuity at about 65 km marks the base of the lunar crust according to Toksöz *et al.*, and below this depth velocities are higher (greater than 8 km/sec).

E. MAGNETIC RESULTS

The Apollo 12, 14, and 15 surface magnetomer experiments have measured magnetic fields ranging from 6 ± 4 gammas to 103 ± 5 gammas (Dyal *et al.*, 1972). The fields measured by the Apollo 16 mission are apparently considerably greater, reaching as high as approximately 300 gammas.

Analysis of data from the Apollo 15 subsatellite magnetometer established that the fields measured on the lunar surface during previous lunar missions are not isolated but part of a complex lunar magnetic field (Coleman *et al.*, 1972).

All breccias and igneous rocks measured to date show remanent magnetism. The intensities are as low as about 2×10^{-6} emu/gm and the mean value is about 5×10^{-6} emu/gm (Strangway *et al.*, 1971). The main component of the magnetism is stable and is unlikely to be an artifact. The fields measured in orbit and at the lunar surface are therefore most likely due to remanent magnetization of the rocks.

The lunar surface magnetometer, the subsatellite magnetometer and the lunar samples themselves all indicate that the lunar surface was exposed to a magnetic field at the time the samples passed through the Curie temperature. Sonett *et al.* (1971) have calculated that the strength of this field must have been at least 1000 gammas. Since the range in age of rocks on which both crystallization age and remanent magnetism have been measured extends from 4.0 to 3.1 b.y. (Strangway, 1972), it appears that the fossil field existed for at least this period of time. If one accepts that the breccias (whose formation ages are unknown, but in some cases are younger than the crystalline rocks) also acquired their remanent magnetism by cooling in the presence of a field, then the field can be shown to have existed even more recently (Strangway *et al.*, 1972).

Strangway *et al.* (1972) point out that the field is of unknown origin and may be explained in one of four ways: (1) a short-lived field was produced by meteorite impact or igneous activity, (2) a field was present in the young solar system, (3) the moon approached the earth close enough in the past for the earth's field to magnetize lunar surface rocks, or (4) the moon once had a molten core which acted as a dynamo.

Runcorn *et al.* (1970), Pearce *et al.* (1971), Runcorn (1972) and Strangway (1972) favor the lunar core hypothesis since the first hypothesis is difficult to justify, and the second and third hypotheses postulate fields that are unlikely to have persisted over the range of time required. Runcorn *et al.* (1970) have shown that an iron core up to 20% of the lunar radius (1.5% of the mass) is permissible within the constraints of the known moments of inertia for the moon.

Brett (1973) points out that a molten iron core is grossly incompatible with the lunar thermal models discussed above and that the existence of such a core would require most of the moon to be at or near melting temperatures (in the earth a molten iron core surrounded by a solid silicate mantle is permissible because the high pressures at the core-mantle boundary greatly elevate the melting point of silicates with respect to iron). Dynamo action in molten pods of iron, as an alternative to a core (Pearce *et al.*, 1972), does little to alleviate the situation. Brett (1973) therefore proposes that if a molten core, pods or layer of conducting material is required to produce a dynamo, then it consists largely of iron alloyed with sulfur in near eutectic proportions. Such a core is molten at about 990°C at the center of the moon. Little change is required in many thermal models if one postulates that the core formed a few hundred million years after the accretion of the moon 4.6 b.y. ago.

The dynamo could have stopped by lowering of the magnetic Reynolds number (a function of electrical conductivity of the material forming the dynamo, the magnetic permeability, the rotational velocity, and the diameter of the dynamo), or by solidification of the core. Solidification now appears unlikely if the silicate interior is partially molten (above 1500°C) below 1000 km, as suggested by the seismic results. It must be stressed, however, that the existence of an internal lunar dynamo at any time in the past is still speculative.

III. Chemical Data From Lunar Orbit

Two experiments deployed from the Service Module during lunar orbit during the Apollo 15 and 16 missions enabled geochemical mapping of a considerable portion of the lunar surface. The x-ray fluorescence experiment (Adler *et al.*, 1972) detected secondary (fluorescent) x-rays excited by radiation of the lunar surface with solar-derived x-rays. Under quiet sun conditions the solar x-ray flux is best suited for exciting the light elements including the abundant elements Si, Al, and Mg. The experiment worked best for measuring the important ratio Al/Si that, when combined with data from the gamma spectrometer experiment and the ground truth established from returned lunar samples, allows approximate identification of lunar rock types on the surface.

Adler *et al.* (1972) show that albedo differences at least partially reflect chemical differences. The x-ray experiment extends and maps the significant chemical differences that were observed between the lunar highlands and maria by the Surveyor program and in previous returned lunar samples. The highlands are Al-rich with respect to the maria, and the low Al values measured over the maria are consistent with the analyses of returned mare basalt samples. There appear to be significant chemical differences from mare to mare and even within individual maria. The spectrometer detected no evidence suggesting large outcrops of Si-rich ("granitic") material on the lunar surface. Significant compositional differences exist within the highlands, the bulk of which appear to lie in the Al/Si range of anorthositic gabbros and gabbroic anorthosites.

The gamma-ray spectrometer experiment (Arnold *et al.*, 1972) carried in lunar orbit on Apollo 14 and 16 examined the energy region from 0.3 to 10 MeV. This region contains major contributions from the radioactive elements Th, U, and K, that are important in understanding both the degree and extent of igneous differentiation and the thermal budget of the moon. The highest values of K, Th, and U in the area traversed are in Mare Imbrium and Oceanus Procellarum. These values may indicate a high concentration of KREEP, a special rock type, whose chemistry is discussed below. There was little or no rise in count rate in the eastern maria compared to the highlands (Metzger *et al.*, 1972). The bulk of the highlands is low in K, Th, and U; however a rise in count rate occurred in the southernmost latitudes traversed on the lunar backside. The K/U ratio is considerably lower than in terrestrial samples in all regions traversed; analysis of returned samples confirms this observation.

IV. Lunar Geology

The returned samples have supplied convincing evidence that the lunar regolith (the soil and pulverized rock layer resting on bedrock) is formed by repeated communition of solid rock by meteoroid impact as suggested by E. M. Shoemaker. The median particle size (by weight) is less than 100 μ. Material in the regolith is commonly from a much larger geographical area than the rocks in any given area of the moon's surface, since finer particles are dispersed more widely than the larger rocks when disturbed by impact. Hinners (1971) presents a rather detailed discussion in which he shows that statistically most rock fragments at the Apollo 11 and 12 sites are locally derived.

For a number of years, the question of whether or not the bulk of lunar craters was formed by impact or volcanism was a subject of considerable controversy, but most workers today agree with Gilbert (1893) that most craters were formed by impact. Since this conclusion has been established, the concept that the age of a surface on the moon is proportional to the density of craters that it contains has received little criticism. This principle, combined with the geologic law of superposition, enabled members of the U.S. Geological Survey to construct detailed maps of the moon's stratigraphy during the 1960's (see compilation by Mutch, 1970). It became possible to state that the heavily cratered non-mare regions predate the maria and to establish the relative ages of the mare surfaces. The crystallization ages that have been determined for some mare areas make it possible to date other mare areas on the basis of relative crater density with some degree of precision (e.g. Soderblom and Lebofsky, 1972).

The crater density of the most heavily cratered highlands area is more than 19 times the density at Mare Tranquillitatis and more than 40 times the density at Oceanus Procellarum (Hartmann, 1970). The crystallization ages of the two maria are 3.7 and 3.3 b.y. respectively. Even if one assumes that the highlands area is as old as 4.6 b.y., one is forced to conclude that the meteoroid flux was considerably higher in the first half billion years of lunar history than it was later. This has led some authors to postulate that the collisional energy provided by the high meteoroid flux early in lunar history provided a considerable source of heat to the outer portion of the moon.

Geologic mapping of the moon by means of photographs, telescopic viewing, lunar photography from NASA's programs, and some of the excellent observations by Apollo Command Module pilots has enabled geologists to recognize the products of a number of processes that occurred on the lunar surface. These include lava flow fronts, impact ejecta indicative of base

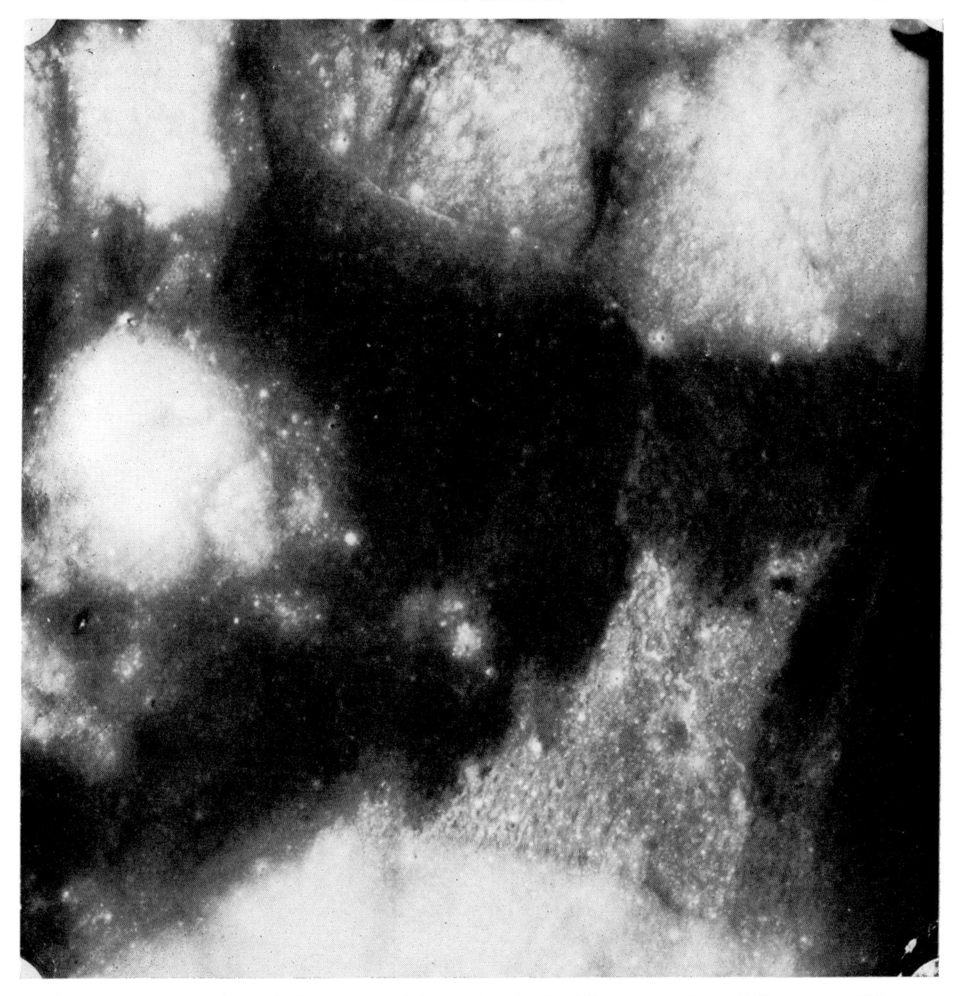

FIGURE 4. Portion of the Taurus Mountains in the Littrow region of the moon. The Apollo 17 landing site is in this area. Note that the high albedo of the upland hills, and the late dark mantling material which may be deposits related to postulated cinder cones in the area.

surge processes, and volcanic cinder cones (Fig. 4). The impact crater density in the ejecta blankets of the cinder cones indicates that they may be considerably younger than some of the maria. Some may be as young as 1 b.y. (Gast, 1972) and may form the last stage of lunar igneous activity. Terrestrial cinder cones form only if considerable gas pressures are present in the magma chambers; this fact suggests that volatile compounds may have been important in such lunar volcanic activity.

Observations on the lunar surface by astronauts supplemented by detailed surface photography have provided information on a number of problems with broad implications. The Apollo 15 mission showed layering in Hadley Rille, which combined with data from previous missions indicates that the mare basins were formed by a series of volcanic flows. Much information has been gained in understanding the processes of erosion and transport of material on the lunar surface. The detailed documentation of samples by astronauts has provided important constraints in interpreting the results of sample analysis.

The present paper will not discuss lunar geology and the evolution of the lunar surface in any detail. The subject is complex but models have not changed drastically in the past 2 years. The interested reader is referred to the reviews of Mutch (1970), Hinners (1971), and Lowman (1972).

V. Lunar Samples

Textural classification indicates that three main types of material have been returned from the moon by the U.S. Apollo program and the Soviet Luna program:

1. Igneous rocks are mostly products of crystallization of lava at the lunar surface, although some of these rocks may have formed in magma chambers below the surface. Some igneous rocks contain considerable glass in their matrices, a fact indicative of rapid cooling. Some rocks contain fairly coarse crystals in a glassy matrix, which suggests that crystallization began below the lunar surface.

2. Regolith fines consist of rock and mineral fragments and angular to rounded glass particles produced by impact processes on the lunar surface. Study of the regolith fines is of importance, not only in understanding lunar erosional processes, but also because material from a wide area of the moon is represented in the fines, whereas most of the larger rock fragments are locally derived. Some workers have used this property to advantage by analysing glasses in the fines in a statistical manner with the object of determining the major element compositions of the most abundant rock types on the lunar surface (e.g. Apollo Soil Survey, 1971; Reid *et al.*, 1972). The method uses the assumption that the statistically preferred compositions are representative of rock types.

3. Breccias consist of angular fragments of crystalline rock, individual minerals, glass, and older breccias in a finer grained matrix of rock and mineral material. They are the products of impact processes on the lunar surface. The breccias found at the mare sites represent indurated or sintered regolith of local origin and are largely products of local impact events. The breccias returned from the Apollo 14 and 15 missions are presumably largely

products of the high energy Imbrian event and represent material ejected from the Imbrium basin when it was formed. Some at least may represent material that never was part of a regolith and that may have been exposed to temperatures over 1000°C (e.g. Williams, 1972) as a result of the Imbrian event. Although the breccias are more complicated texturally than rocks of igneous origin, in many samples more information can be gained from them than from igneous rocks since they commonly contain a variety of rock types.

A. MINERALOGY

The most abundant minerals in lunar material are also abundant in terrestrial basalts and gabbros. They include plagioclase, clinopyroxene, orthopyroxene, olivine, ilmenite, and (Cr, Al, Ti, Mg) spinels (Fig. 5). Over 15 other minerals have been identified; the interested reader is referred to a listing by Frondel (1972) and the summaries of lunar sample investigations by Levinson and Taylor (1971) and Mason and Melson (1970). No hydrous minerals have been positively identified in any rock with the exception of rare goethite (FeOOH) (Agrell *et al.*, 1972). All lunar rocks contain small amounts of metallic Fe or Fe–Ni. The presence of native iron indicate that lunar rocks crystallized under considerably lower partial pressures of H_2O and O_2 than terrestrial basalts (e.g. Wellman, 1969; Brett *et al.*, 1971; Haggerty, 1971; Charles *et al.*, 1971).

No minerals indicative of crystallization under high pressures have so far been found, suggesting that all returned lunar material crystallized no deeper than a few tens of kilometers below the lunar surface.

B. MATERIAL OF MARE ORIGIN

Mare basalts have been returned by the Apollo 11, 12, and 15 missions. Mare-derived material has been found also in the regolith fines returned by the Apollo 14 and Luna 16 missions. Understanding of the mare-forming process is important since the maria occupy some 20% of the lunar front side. Relative ages by the crater density method indicate that maria are the youngest major features on the lunar surface, and albedo differences, orbital geochemical experiments, and the returned samples themselves show that mare material differs chemically from material of non-mare origin.

The rocks texturally resemble terrestrial basalts and appear to have formed by rapid, near-surface crystallization of lavas (e.g. James and Jackson, 1970). In some rocks gas cavities (vesicles and vugs) are abundant (Fig. 6) and indicate the presence of a volatile phase (Wellman, 1969) during crystallization. The rocks range in composition from mare to mare and there is strong evidence, based on chemical and textural data and crater stratigraphy, that the astronauts and the Luna 16 vehicle at Mare Fecunditatis sampled several

FIGURE 5. Thin section of an Apollo 14 igneous rock. The grey laths are plagioclase, the black areas are largely ilmenite, and the most abundant mineral is clinopyroxene.

FIGURE 6. A highly vesicular basalt (15016) returned from the Apollo 15 mission.

flows of different composition at each mare site visited (e.g. James and Jackson, 1970; Warner, 1971; Jakes *et al.*, 1971). Evidence of chemical variation arising from extensive fractional crystallization within thick flows, sills, or dikes is abundant in rocks from Oceanus Procellarum.

Chemically, all returned mare basalts are similar enough to form an easily definable group (Table 1). The differences in chemistry in materials from different maria may partly result from different degrees of partial melting in a homogeneous source region. The major element chemistry of the mare basalts is fairly similar to that of some terrestrial basalts (except that Fe/(Fe + Mg) is higher than that of most terrestrial basalts and Ti in Apollo 11 basalts is higher than in any terrestrial basalt).

The trace and minor element chemistry of the mare basalts is unique so they can readily be distinguished from terrestrial and meteoritic material. The mare basalts are strongly depleted in geochemically volatile elements

Table 1
Chemical composition of major lunar rock types

	Mare Basalts				KREEP		Plagioclase Rich Rocks		Silica Rich Rocks
	XI[1]	XII[2]	XV[3]	L16[4]	XII[5]	XIV[6]	15415[7]	Anortho.[8] Gabbro	Average[9]
SiO_2	40.70	44.95	46.07	43.8	47.6	48.0	44.08	45.3	75
TiO_2	11.00	3.32	2.13	4.9	1.7	2.1	0.02	0.48	16
Al_2O_3	9.43	9.19	8.95	13.65	17.21	17.1	35.49	26.3	14
Cr_2O_3	0.32	0.51	—	0.55	0.17	—	0.20	~0.1	—
FeO	17.42	20.53	21.19	19.35	9.4	10.5	0.23	5.3	1.8
MgO	7.34	9.83	9.51	7.05	9.1	8.7	0.09	7.3	0.3
CaO	10.52	10.94	10.21	10.4	10.2	10.7	19.68	14.6	1.4
Na_2O	0.49	0.28	0.26	0.38	1.0	0.7	0.34	0.3	0.9
K_2O	0.18	0.058	0.034	0.15	0.7	0.5	0.01	0.07	6.8
P_2O_5	0.12	0.088	0.07	—	1.0	—	—	—	—
MnO	0.23	0.27	0.28	0.20	—	—	—	—	0.8

Compilation largely from Gast (1972). 1. Average of 13 analyses reported in Proceedings of the Apollo 11 Lunar Science Conference. 2. Average of 17 analyses reported in Proceedings of the Second Lunar Science Conference. 3. Average of 7 analyses reported by LSPET, 1972. 4. Composition of one crystalline fragment, Vinogradov, 1971. 5. Average of 29 fragments, broad beam microprobe analyses given by Keil *et al.*, 1971; Meyer *et al.*, 1971; Smith *et al.*, 1970. 6. Average of Type B and C glasses, Apollo Soil Survey, 1971. 7. Composition of Rock 15415, a plagioclase-rich rock from Apollo 15 site, LSPET, 1972. 8. Average of high alumina feldspathic basalts, Reid *et al.*, 1972. 9. Average of 4 granitic fragments, Meyer, 1972.

with respect to chondritic abundances. These elements include most of those in groups 1A, 2B, 3A, 4A, 5A, 6A, and 7A of the Periodic Table. The basalts are also depleted in the more noble elements Os, Ir, Au, and Pd as well as in Co, Cu, Ag, and Ni. These are so-called siderophile elements that alloy readily with metallic iron. The mare basalts tend to be enriched in the more refractory elements Li, Be, Ba, Sr, Ca, La, Y, Sc, Ti, Hf, Th, Zr, Ta, Nb, U, and the rare earth elements.

The depletion in volatile elements, which is also found in non-mare material, indicates that at least the outer portion of the moon, which yielded mare basalts by partial melting, is depleted in volatiles since many of these elements are not fractionated into refractory phases during partial melting processes, nor do many of them tend to be enriched in any metallic phase. Biggar *et al.* (1971) claim that the bulk of the volatile elements was lost when lava pools cooled under the high vacuum conditions of the lunar surface. Gibson and Hubbard (1972) have demonstrated experimentally that such volatile loss occurs, but they point out that pressures at depths greater than a fraction of a millimeter in the melt would prevent volatile escape. They therefore suggest that volatile outgassing from lava flows was quantitatively of little significance.

The depletion in the siderophile elements has led some authors (e.g. Ganapathy *et al.*, 1970) to postulate that they were enriched in a metallic phase in the lunar interior. An alternative hypothesis is that the outer portion of the moon was initially depleted in these elements when it accreted.

The enrichment in refractory elements with respect to chondrites is certainly at least in part due to the fact that these elements are preferentially enriched in the melt in partial melting processes. However, some elements, like the rare earths, are so greatly enriched in mare basalts (Fig. 7) that some authors (e.g. Gast, 1972) have concluded that the outer portion of the moon was

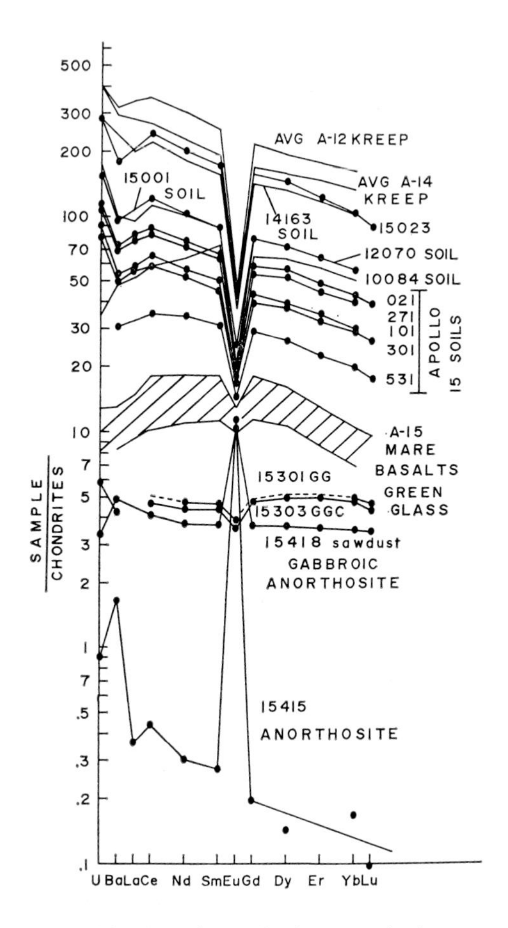

FIGURE 7. Plot (after Hubbard *et al.*, 1972) of rare earth elements and some large ion lithophile elements normalized against chondritic abundances. Note that the negative Eu anomaly in mare basalts and KREEP and the positive anomaly in anorthosite and anorthositic gabbro.

initially enriched in these elements when it accreted. An alternative hypothesis is that the bulk of the moon was molten in the past and that these elements were highly concentrated in the lunar crust by repeated partial melting. This hypothesis is difficult to fit within the constraints presently imposed by our knowledge of the moon's thermal history.

Mare basalts, in fact lunar material generally, are characterized by low K/U ratios with respect to terrestrial rocks and chondritic meteorites (Fig. 8).

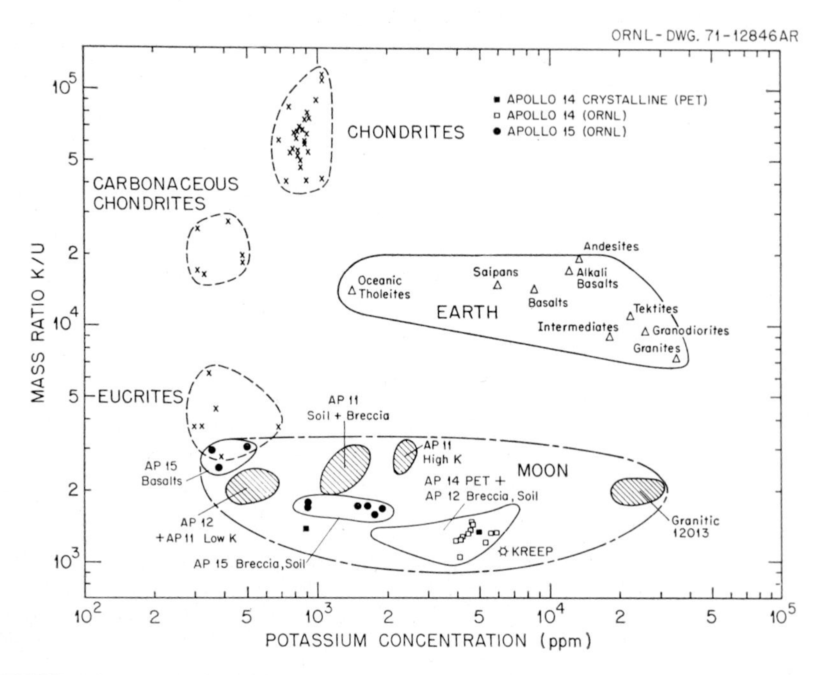

FIGURE 8. Mass ratio K/U versus K concentration for terrestrial rocks, meteorites and lunar rocks. Note that lunar rocks do not overlap terrestrial rocks. This is a strong argument against the fission hypothesis. Note that tektites, for which a lunar origin has been proposed by some, lie in the terrestrial range (Figure is from O'Kelley *et al.*, 1972)

The ratio in lunar rocks is relatively constant, which reinforces the conclusions that the outer portion of the moon is depleted in K, and that the role of U in the thermal budget of the moon is more important than in that of the earth. The ratio $^{238}U/^{204}Pb$ (primordial Pb) is high compared to that in the earth and chondrites and leads to the conclusion that Pb, a volatile element, is depleted in the moon, or that Pb, which is also a chalcophile element (readily combines with S), is enriched in Fe–FeS rich bodies below the lunar surface (Murthy *et al.*, 1971; Brett, 1973).

Mare basalts are strongly depleted in europium with respect to the other rare earth elements (Fig. 7). This so-called Eu anomaly is probably due to the reducing conditions prevalent during partial melting in the lunar interior. Europium, unlike the other rare earth elements, is concentrated in plagioclase in lunar rocks.

A number of radiometric age determination methods are used to date lunar material; this paper will not discuss the methods and controversies associated with them. Wetherill (1971) ably discusses some of these complex problems. The mare basalts range in crystallization age from about 3.3 b.y. (Oceanus Procellarum, Palus Putredinis) to 3.7 b.y. (Mare Tranquillitatis). Combined with determination of relative ages by the crater density method, the radiometric ages indicate that the mare-forming process entended from a little less than 4 b.y. ago to a little more than 3 b.y. ago (e.g. Soderblom and Lebrofsky, 1972).

The density of mare basalts is close to that of the bulk moon (3.35 gm/cc). Ringwood and Essene (1970) have shown that mare basalts would transform at the pressures and temperatures of the lunar interior to a mineral assemblage whose density (c. 3.7 gm/cc) is greater than that of the bulk moon. Therefore, in agreement with Wetherill (1968), they claim that the moon cannot be basaltic throughout. It is clear on the basis of chemistry, texture, distribution and geologic features on the mare surfaces that the mare basalts are products of partial melting within the lunar interior (e.g. Ringwood, 1970a).

No matter what current theory for the origin of mascons is accepted, it is clear that at the time of extrusion of the mare basalts, the exterior of the moon was sufficiently rigid to support mascons, which requires a lunar lithosphere at least 150 km thick (Kaula, 1969). The source region for the mare basalts must therefore have been below that depth. Also, Ringwood and Essene (1970) have shown experimentally that if melts of mare basalt composition were derived from depths greater than about 500 km, the partial melt would be in equilibrium with clinopyroxene and garnet. Such a source rock would have a density greater than that of the bulk moon and would have to occupy a small volume of the bulk moon to satisfy the lunar moments of inertia. Ringwood and Essene therefore propose that the mare basalts were most likely derived from depths of 200 to 500 km.

Ringwood and Essene (1970) show that at lunar depths between 200 and 500 km the mare basalts would be in equilibrium with a pyroxene-rich source rock, and thus propose that the mare basalts are products of partial melting of a pyroxenite or olivine pyroxenite. Toksöz *et al.* (1972) postulate the same composition for the upper lunar mantle on the basis of seismic studies.

On the other hand, the large negative europium anomaly in mare basalts suggests that they originated in a source area that contained plagioclase (e.g.

Gast, 1972). Ringwood and Essene (1970) argue that significant amounts of plagioclase cannot exist at the depths postulated. An alternative is that the primitive source rock itself has a negative Eu anomaly (Papanastassiou and Wasserburg, 1971). One might suggest that since some plagioclase rich non-mare samples have a strong positive Eu anomaly (Hubbard and Gast, 1971) that these represent a complementary differentiate to the mare basalts, but from phase equilibria and trace element considerations this hypothesis is difficult to justify.

To summarize, although the exact composition of the source of mare basalts is presently uncertain, the basalts probably originated at depths between 200–500 km and from a pyroxenitic parent rock (Table 2), a composition which is consistent with the seismic data. The origin of the Eu anomaly remains a mystery due to the probable absence of significant plagioclase below 200 km in the lunar mantle.

Table 2
Model compositions of the lunar interior

	1	2	3	4
SiO_2	53.1	49.1	48.0	45.2
TiO_2	1.0	0.45	0.8	0.4
Al_2O_3	5.0	6.0	16.0	7.6
FeO	13.5	18.0	9.0	19.7
Cr_2O_3	0.4	0.5	—	0.43
MgO	22.5	19.6	14.0	17.9
CaO	4.0	5.3	12.0	8.1
Na_2O	0.1	0.1	—	0.1

1. Lunar Interior, Ringwood and Essene, 1970. 2. Lower Lunar Mantle (Gast, 1972). 3. Mean Composition of the Upper 150 kilometers (Gast, 1972). 4. Homogeneous green glass (Ridley *et al.*, 1972).

C. MATERIAL OF NON-MARE ORIGIN

The soil of the Apollo 11 return provided evidence of lunar material of non-mare origin, and other samples have been returned from each subsequent mission. An understanding of the petrogenesis and evolution of the non-mare rocks is extremely important because material of non-mare origin occupies over 80% of the lunar surface and represents a record of the very early history of the moon, since endogenic processes affecting the lunar highlands were virtually complete some 4 b.y. ago.

The material of non-mare origin can be subdivided into three main types: (1) material of anorthositic affinity, (2) a rock type with high abundances of K, rare earth elements, and P (named "KREEP"), and (3) siliceous material containing high K that although termed granitic is somewhat different from terrestrial granitic material.

Material of anorthositic (rich in calcic plagioclase) affinity was identified by a number of workers in the Apollo 11 soil (e.g. Wood, 1970). Similar material was identified in the Luna 16, Apollo 12, and Apollo 14 soils. At the Hadley-Apennine site visited by the Apollo 15 crew, this material is abundant in the soil, especially that from the Apennine Front. Several rocks were recovered there that are largely anorthositic in composition, and clasts of anorthositic affinity are fairly abundant in breccias. (LSPET, 1972).

The material ranges from truly anorthositic through gabbroic anorthosite to anorthositic gabbro. Reid *et al.* (1972) statistically surveyed compositions of glass soil particles from a number of landing sites and found that the bulk of material of anorthositic affinity has a fairly narrow range of composition corresponding to that of anorthositic gabbro (a rock rich in plagioclase and pyroxene) (Table 1). They postulate that rock of this composition is abundant in the lunar highlands. Data from the orbital x-ray fluorescence experiment (Adler *et al.*, 1972) are consistent with this conclusion.

The anorthositic gabbros and anorthosites (incorrect terms, both texturally and to a lesser extent compositionally, with respect to terrestrial rocks of those names) are richer in Ca and Al than mare basalts (Fig. 9), and poorer in rare earth elements though still enriched with respect to chondrites. They have positive Eu anomalies (Fig. 7). Elements of large ionic radius which tend to be enriched in the liquid phase during partial melting (e.g. K, Ba, Zr), termed large ion lithophile elements, are depleted in the anorthosites and anorthositic gabbros. Hubbard *et al.* (1971), on the basis of trace element chemistry of the anorthositic clan, postulate that there are at least two types of anorthositic material, one derived from a source rich in the above elements (KREEP), the other from a source poor in these elements. Textures in anorthositic rocks and clasts returned by Apollo 15 are characteristically brecciated, although some rocks and soil particles retain textural evidence of crystals in an igneous melt indicative of gravitational accumulation either by sinking or flotation (so-called cumulate textures).

The material termed KREEP was first identified as an abundant component in the soil at the Apollo 12 (Mare Procellarum) site (Meyer *et al.*, 1971). Since the landing site lies on a ray from the crater Copernicus, Meyer and others propose that KREEP was ejected from Copernicus. Some workers (e.g. Wasson and Baedecker, 1972) disagree with this conclusion. KREEP material was identified as the predominant material at the Apollo 14 (Fra Mauro) site (LSPET, 1971), as predicted by a number of authors, e.g. Meyer

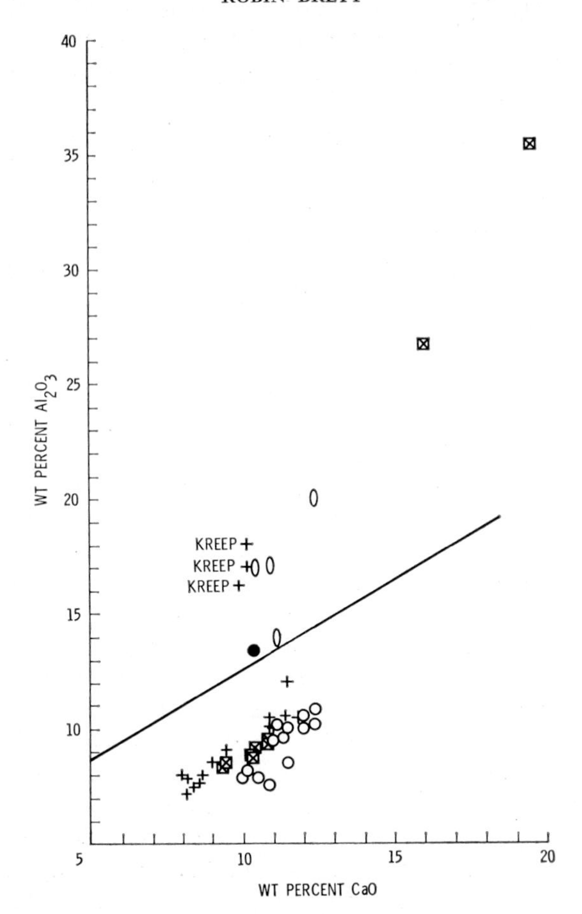

FIGURE 9. Plot of CaO versus Al_2O_3 for rocks from Apollo 11, 12, 14, and 15, and Luna 16. Note that rocks of proposed non-mare origin lie on the high Al_2O_3 side of the line; rocks of mare origin are on the low Al_2O_3 side. All stony meteorites lie on the line. The bulk of the points are from the plot of Gast (1972).

et al. (1971). KREEP also occurs in the Apollo 15 sample return (Meyer, 1972). On the basis of orbital γ-ray data, Metzger *et al.* (1972) postulate that the surfaces of Oceanus Procellarum and Mare Imbrium are relatively rich in the KREEP component.

The KREEP component is basaltic in composition (Table 1), its closest terrestrial equivalent being a high alumina tholeiite. It is richer in Al, Zr, P, rare earth elements, and Ba than the mare basalts, and poorer in Fe and Ti. It shows a more pronounced negative Eu anomaly than the mare basalts (Fig. 7).

The bulk of KREEP material is brecciated and much of the material found in the Apollo 12 soil is glassy. Crystalline examples of KREEP were found in Apollo 14 (LSPET, 1971) and 15 (Meyer, 1972) returns.

The Fra Mauro formation, sampled by the Apollo 14 mission, represents material excavated from the Imbrium basin (e.g. Mutch, 1970). Fra Mauro material was excavated from the crater Copernicus (Schmitt *et al.*, 1967). The orbital γ-ray data indicate the greatest concentration of the KREEP component in the vicinity of the Imbrium basin. It therefore appears likely, as pointed out by Hinners (1971), that the original excavation of KREEP was caused by the Imbrian event.

Schonfeld (1972) calculates that the original crystallization age of KREEP material is about 4.4 b.y. The Rb–Sr systematics of rock 12013 (Lunatic Asylum, 1970) of which KREEP is an abundant component (Meyer *et al.*, 1971) indicate that KREEP material crystallized before 4.0 b.y. KREEP is the most radiogenic component in the lunar soils, and therefore contributes significantly to, and possibly dominates the Rb–Sr, U–Th–Pb, and K–Ar systematics of the soil.

Fragments of granitic composition occur in very small amounts in soils from most lunar landing sites. They contain quartz, plagioclase, K feldspar, and an iron rich pyroxene. Textural evidence indicates that the pyroxene cooled slowly (Meyer, 1972). An average composition is listed in Table 1. The crystallization age of the granitic component is unknown; however, the Lunatic Asylum (1970), from studies of the granitic component of rock 12013, has shown that original crystallization occurred before 4.0 b.y.

The origin of the granitic component is obscure. Meyer (1972) suggests that it possibly formed by fractional crystallization and filter pressing of magma of KREEP composition, or by partial melting during thermal metamorphism of KREEP material. Hubbard *et al.* (1970) suggest that it possibly formed by filter pressing of an immiscible granitic phase that occurs in mare basalts (Roedder and Weiblen, 1971).

Ridley *et al.* (1972) have recently discovered an extremely homogeneous green glass, rich in Fe and Mg and low in Ti (Table 2) that occurs at the Apollo 15 and other landing sites. The glass corresponds to a plagioclase-bearing ultramafic rock rich in olivine, orthopyroxene, and clinopyroxene. They point out that the composition of the rock is similar to that suggested by Gast (1972) for the outer lunar mantle (Table 2).

It is clear from the above discussion and the orbital x-ray fluorescence and γ-ray data that rocks of anorthositic affinity are the most abundant type in the non-mare areas of the moon. Of lesser abundance is the KREEP component. Little is known of the crystallization ages of the anorthositic rocks; however, data by Turner (1972) indicate that an anorthositic rock recovered by the Apollo 15 mission originally crystallized more than 4.1 b.y. ago.

Wood *et al.* (1970) suggest that the anorthositic highlands were formed by plagioclase-rich material floating to the surface of a melted outer moon. A number of other workers have suggested variations on this basic theory, which is attractive for its simplicity. Unfortunately, the origin of terrestrial anorthositic rocks is little understood and there are few example of plagioclase floating in terrestrial melts. It is also difficult, if not impossible, to derive the anorthositic rocks as cumulates and the later mare basalts from the same source rock, since the anorthositic rocks must be derived from a Ca and Al rich source (e.g. Gast, 1972), and, as discussed earlier, it is difficult to justify high Ca and Al in the lunar mantle.

The KREEP basalts cannot be derived from the same source region as the mare basalts and Hubbard and Gast (1971) suggest that they represent a small degree of partial melting of a shallow, primitive layer less than 100 km deep that was substantially enriched relative to chondritic meteorites in Ca, Al and refractory trace elements. Some anorthositic material could have been derived from the KREEP magma; more extensive melting of the same source area may have generated the melts from which the low K anorthosites separated by fractional crystallization (Hubbard *et al.*, 1971).

VI. Lunar Evolution

The commonly accepted formation age of the earth and all meteorites but one is approximately 4.6 b.y. Papanastassiou *et al.* (1970) show, on the basis of Rb–Sr systematics, that the material from which the moon accreted could not have resided in the solar nebula more than about 4 m.y. longer than the basaltic achondrites, which formed about 4.6 b.y. ago. They also suggest that Rb–Sr systematics in mare basalts are consistent with their having remained a closed system for 4.6 b.y. Whole rock Rb–Sr ages and model (apparent) ages for lunar regolith samples are commonly 4.6 b.y. (e.g. Cliff *et al.*, 1972). Arguments based on Rb–Sr systematics place the formation of the moon at 4.6 b.y. and require that the material which formed the moon was isolated from the solar nebula for no more than 100–200 m.y. (Papanastassiou and Wasserburg, 1971). The accumulated data thus strongly imply that the moon was formed at about the same time as the other bodies in the solar system for which cosmochronological data are available.

Reasonably good data are available at present on the composition of the solar corona and, by inference, the solar nebula. For some years many workers assumed that the bulk composition of the earth is that of the chondritic meteorites, which are believed to represent the most primitive examples of condensates from the solar nebula. More recently it has become clear that there are some differences between the earth's composition and that of chondrites, particularly in regard to an apparent depletion in alkali metals

(Gast, 1960) and apparent enrichment in refractory elements in the earth (Gast, 1968). Gast (1972) has examined the problem with regard to the moon. He assumes that one tenth of the outer crust of the moon is of the composition of the KREEP component. Calculations show that to obtain the enrichment in elements such as the light rare earth elements, U and Ba, a shell of chondritic composition 300 km thick (40% of the mass of the moon) must have had these elements totally extracted from it. It is unlikely that the moon underwent such large scale melting, even making the extremely unlikely assumption that extraction was 100% efficient. On the other hand, the abundance of KREEP in the lunar crust has not been established, and if its abundance in the lunar crust is an order of magnitude less than that assumed by Gast, then derivation of the outer portion of the moon from a chondritic moon is more plausible, but still very unlikely.

If one accepts the unlikely conclusion that the outer 300 km of the moon was depleted in certain elements to enrich the crust, then one has difficulty in explaining the considerable enrichment of the same elements in mare basalts, which are believed to be derived from depths between 200–500 km. On the other hand, if one accepts that the mare basalts were derived from a previously unmelted source with the composition of the bulk moon, then one is forced to conclude that the basalts were not derived from a chondritic source, but rather a source enriched in large ion lithophile elements (e.g. Gast, 1972).

There appear to be only two solutions to the dilemma; either to agree with Gast (1972) that the moon is not chondritic in composition and that it was accretionally zoned or to propose that the bulk of the moon was molten early in its history, causing enrichment in certain elements near the surface. As mentioned earlier, there are immense difficulties involved in providing a heat source for large scale melting at depth early in lunar history.

Gast and Hubbard (1970) point out that the low Rb/Sr and K/U ratios of lunar rocks suggest that their source was enriched in refractory elements like Sr and U and depleted in volatile elements like Rb and K with respect to chondrites. Hubbard and Gast (1971) suggest that chemical and isotopic characteristics of small lunar fragments of non-mare origin require liquids produced by partial melting 4.3 to 4.4 b.y. ago. Nyquist *et al.* (1972) point out that non-mare materials with crystallization ages of 3.9 to 4.0 b.y. may be derived from rocks crystallized much earlier in lunar history. Gast (1972) states that whole rock Rb–Sr ages, particularly on KREEP basalts, indicate that partial melting of silicate liquids started within 200 m.y. of the time of the moon's formation. Wasserburg *et al.* (1972) also propose that an igneous crust rich in radioactive elements existed close to 4.6 b.y. ago.

The chemical composition of the non-mare materials indicates that they were differentiated from a source richer in Al and poorer in Fe than the source of the mare basalts. Gast (1972) therefore proposes that the outer portion of the moon was condensed from a hot solar nebula that was depleted in volatile elements and enriched in refractory elements including Al and U. Gast points out that the variation in chemistry in mare basalts not only requires a range in the degree of partial melting (from a minimum of 3 to 10% to a maximum of 20 to 30%), but also a range in depth of melting coupled with a variation in the mineralogy and chemistry of the moon with depth. Most workers agree with these interpretations, but some believe that the heterogeneity of lunar composition originated not by accretional zoning, but by large scale internal differentiation which requires an extensive fraction of the moon to have been molten in very early lunar history. No coherent description of this hypothesis has yet been published, however, and its advocates must provide a heat source for the large scale melting and explain how the moon cooled to its present temperature.

The following lunar events are tentatively proposed from the data cited above:

1. Heterogeneous accretion of the moon about 4.6 b.y. ago with an interior rich in pyroxene and an outer portion (at least the outer 50 km) enriched in Al and large ion lithophile elements (including U, Th, and K—Papanastassiou and Wasserburg, 1971) and depleted in Fe (Gast, 1972).

2. Early melting of the outer 150 km of the moon and differentiation to produce a suite of rocks ranging from anorthosite to gabbro to KREEP (Gast, 1972). These rocks now comprise the upland (non-mare) portions of the moon. The KREEP component represents a very small degree of partial melting of the outer Al rich zone (Hubbard and Gast, 1971) while the anorthosites are a product of gravitational differentiation (e.g. Wood *et al.*, 1970). This stage of upland differentiation lasted until at least 4.4 b.y. ago (e.g. Schonfeld, 1972) forming an anorthosite-rich crust about 65 km thick (e.g. Toksöz *et al.*, 1972).

3. From 4.6 b.y. until about 4.0 b.y. age the moon was being much more intensely bombarded than in its subsequent history; the highlands were heavily cratered, and finally most circular mare basins were excavated. The Imbrium basin, the last large mare basin to form but one was formed about 3.9 b.y. ago, because rocks from the Fra Mauro site record a recrystallization age at the time of the Imbrian event or earlier possess this age (e.g. Wasserburg *et al.*, 1972).

4. Before 4.1 b.y. a conductive molten core of Fe + FeS may have formed (Brett, 1973) if the remanent magnetism observed in the lunar samples was caused by an internal lunar dynamo.

5. The zone of melting extended deeper with time, as the effect of radioactive heating predominated over accretional heating, (Gast and McConnell, 1972), causing partial melting of the pyroxene rich layer at depths of 200–500 km (Ringwood and Essene, 1970), during the time period extending from

about 3.2 to 3.7 b.y. this melting yielded the mare basalts that flooded the mare basins and other low areas confined almost exclusively to the lunar front side (e.g. Wasserburg *et al.*, 1972). The outer portion of the moon had become sufficiently rigid at this stage to support the mascons. Internally derived igneous activity virtually ceased about 3 b.y. ago, due largely to a decrease in concentration of radioactive elements with depth within the moon (Gast, 1972; Wasserburg *et al.*, 1972). An alternative explanation might be that convection cells were operating in the interior (e.g. Tozer, 1972) thus buffering any increase of temperature in the lunar interior.

6. Little activity of internal origin happened to the surface of the moon after 3 b.y. Activity was restricted to cratering at a much lower rate than previously, seismic events, and small scale igneous activity possibly producing mantling material of low albedo and cinder cones. Some of these cones and dark mantles post-date the relatively recent Copernican impact (Schmitt *et al.*, 1967).

VII. Origin of the Moon

The present hypotheses on the origin of the moon are: (1) the moon was formed by fission from the earth, (2) the moon was captured by the earth, (3) the moon was formed as a binary planet to the earth, and (4) the moon formed from a sediment ring of earth-orbiting planetesimals.

The fission theory was first suggested by Sir George Darwin who, in 1878, proposed that the moon was separated from the earth by the action of tidal forces. The theory was recently revived by Wise (1969) and O'Keefe (1969). There are strong physical arguments against this origin, e.g. Jeffreys (1952), and now that lunar samples are available, the chemical arguments are even stronger (Gast, 1972). The differences in Fe/Mg, K/U, and $^{238}U/^{204}Pb$ ratios and the abundances of siderophile elements in the earth and moon argue against a common origin for the earth and moon. Almost all workers now reject this hypothesis for lunar origin.

The capture hypothesis has enjoyed popularity partly because it is restricted almost entirely to a problem of celestial mechanics to which chemical data can contribute little. This and the other hypotheses are well discussed by Kaula (1971b). The hypothesis argues that the moon accreted elsewhere in the solar system and was captured by the earth. The main problems, as Urey and McDonald (1971) point out, are dissipation of the energy of capture and change from the original orbit to the present nearly circular orbit. Gersternkorn (1955, 1957) proposes that the moon was captured in retrograde orbit and that the energy of capture was dissipated by tidal action in both the earth and moon. McDonald (1964) calculated that such a capture would have taken place 1.79 b.y. ago, which from geological evidence is impossible. Gersternkorn (1969) revised his earlier calculations but his revised hypothesis places the moon as close to the earth as $1.5R_E$. This is well inside the Roche limit where the moon would break up. The hypothesis is therefore highly improbable.

Singer (1968) proposes a mechanically more feasible capture hypothesis than that of Gersternkorn; it has the added advantage that capture could take place close to 4.6 b.y. ago (Singer, 1970). The hypothesis requires that the moon approach the earth at least as close as $2.6R_E$ (Roche limit is $2.89R_E$). Urey and MacDonald (1971) suggest capture at $30R_E$, with addition of material in geocentric orbit amounting to some 20% of the moon's mass. The main problem with all capture hypotheses is that the probability of capture is very low.

The simplified binary planet hypothesis suffers from two major problems. First, if the moon formed as a twin planet to the earth, those problems involving the differences in chemistry between the earth and the moon are encountered. Secondly, the inclination of the moon's orbit to the earth's equator presents a problem. This angle presently varies between 18° and 28°. Extrapolation back through time indicates that this angle seldom approached zero, except when the moon was at $18R_E$ (Goldreich, 1966). An origin as a binary planet from a ring of planetesimals, or by fission of the moon from the earth requires that the moon lie in a near equatorial orbit and remain there (e.g. Urey and MacDonald, 1971). Catastrophic changing of this orbital angle by asteroid impact requires an impacting body of almost the size of the moon. There would surely be evidence of this impact today had it occurred (S. F. Singer, public lecture).

A modified version of the binary planet hypothesis has been presented by Ruskol (1960, 1963) and Öpik (1961). They propose that the moon accreted from a sediment ring somewhat analogous to the rings of Saturn, only more massive. This theory has been developed by Ringwood (e.g. 1970b, 1972), Cameron (1966, 1970), and Berlage (1967). Ringwood's argument, which is largely geochemical, requires that during the later stages of accretion of the earth, a massive primitive atmosphere developed by evaporation of silicate materials accreting on the earth. This atmosphere was driven off as the sun passed through a T-Tauri phase. The nonvolatile elements condensed to a swarm of planetesimals which later accreted to form the moon, as advocated by Öpik, the more volatile elements escaped from earth orbit. The hypothesis best explains the observed lunar chemistry. The main objections to the hypothesis are (1) that it has neither been demonstrated that a T-Tauri sun can remove an atmosphere of high mean atomic number, nor that the sun passed through a T-Tauri stage, and (2) the problem of the moon's orbital inclination with respect to the earth's equator. Ringwood (1972) points out that Goldreich's theory on orbital inclination may not apply if a large proportion of the moon's mass possesses low viscosity, and that the non-zero inclination may have been caused by non-symmetrical escape of the primitive atmosphere or by early impact.

It is clear that the origin of the moon is unknown. The two most likely theories are the sediment-ring theory and the capture theory. The probability of the former is unknown; the latter has low probability. It is considerably easier to unravel the evolution of a planet than to determine its origin. The Apollo and Luna programs have refined thoughts on the origin of the moon somewhat, but it is possible that we may never have a definitive answer to this major question.

Conclusions

Our understanding of the moon has advanced rapidly in the last three years. Never before have so many scientists of different disciplines worked so strenuously on a single problem. The impact of the lunar sample program and experiments on our understanding of the moon has been immense. In the next five years we should have considerably more knowledge of the bulk chemistry of the moon, its internal structure, the partial melting processes that took place, the evolution of lunar magmas, and the thermal history of the moon. This knowledge will have important application to understanding the early history of our own planet, but the ultimate origin of the moon and the earth may still remain a mystery.

Acknowledgements

Owing to delays the above paper includes results obtained only up to May, 1972. Considerable advances have been made since then which I have been unable to add to the text, with the exception of the important new seismic and heat flow data and their implications concerning the lunar thermal regime. In the main, the new data obtained tend to confirm and add detail to the summary presented here.

I thank M. N. Bass, Abigail Brett, E. K. Gibson, Jr., N. W. Hinners, Charles Meyer, Jr., H. H. Schmitt, and R. J. Williams for careful reviews. Inaccuracies and omissions are all mine, however. I thank P. W. Gast, N. J. Hubbard, R. J. McConnell, and G. D. O'Kelley for providing unpublished figures.

References

Adler, I., Trombka, J., Gerard, J., Schmadebeck, R., Lowman, P., Blodgett, H., Yin, L., Eller, E., Lamothe, R., Gorenstein, P., Bjorkholm, P., Harris, B., and Hursky, H. (1972). *In* Apollo 15 Preliminary Science Rept., NASA SP-289.

Agrell, S. O., Scoon, J. H., Long, J. V. P., and Coles, J. N. (1972). *In* Lunar Science-III. Lunar Science Inst. Contr. No. 88.
Anderson, D. L. and Phinney, R. A. (1967). *In* "Mantles of the Earth and Terrestrial Planets" (S. K. Runcorn, ed.). *Interscience.*
Apollo Soil Survey (1971). *Earth Planet. Sci. Lett.* **12,** 49–54.
Arnold, J. R., Peterson, L. E., Metzger, A. E., and Trombka, J. I. (1972). *In* Apollo 15 Preliminary Science Rept., NASA SP-289.
Berlage, H. P. (1967). *In* "Mantles of the Earth and Terrestrial Planets" (S. K. Runcorn, ed.). J. Wiley, New York.
Biggar, G. M., O'Hara, M. J., Peckett, A., and Humphries, D. J. (1971). *Geochim. Cosmochim. Acta, Suppl.* **2,** 617–644.
Brett, R. (1973). *Geochim. Cosmochim. Acta* **37,** 165–170.
Brett, R., Butler, P., Jr., Meyer, C., Jr., Reid, A. M., Takeda, H., and Williams, R. J. (1971). *Geochim. Cosmochim. Acta Suppl.* **2,** 301–317.
Cameron, A. G. W. (1966). *In* "The Earth-Moon System" (B. Marsden and A. G. W. Cameron, eds.).
Cameron, A. G. W. (1970). *Trans. Amer. Geophys. Union* **51,** 728.
Charles, R. W., Hewitt, D. A., Wones, D. R. (1971). *Geochim. Cosmochim. Acta Suppl.* **2,** 645–664.
Cliff, R. A., Lee-Hu, C., and Wetherill, G. W. (1972). *J. Geophys. Res.* **77,** 2007–2013.
Coleman, P. J., Jr., Schubert, G., Russell, C. T., and Sharp, L. R. (1972). *In* "Apollo 15 Preliminary Science Rept.", NASA SP-289.
Dyal, P., and Parkin, C. W. (1972). *The Moon* **4,** 63–87.
Dyal, P., Parkin. C. W., and Sonnett, C. P. (1972). *In* Apollo 15 Preliminary Science Rept., NASA SP-289.
Fricker, P. E., Reynolds, R. T., and Simmons, A. L. (1967). *J. Geophys. Res.* **72,** 2649.
Frondel, J. W. (1972). Preprint.
Ganapathy, R., Keays, R. R., Laul, J. C., and Anders, E. (1970). *Geochim. Cosmochim. Acta Suppl.* **1,** 1117–1142.
Gast, P. W. (1960). *J. Geophys. Res.* **65,** 1287–1297.
Gast, P. W. (1968). *In* "History of the Earth's Crust" (R. Phinney, ed.). Princeton Univ. Press.
Gast, P. W. (1972). *The Moon*, **5,** 121–148.
Gast, P. W. and Hubbard, N. J. (1970). *Science, N.Y.* **167,** 485–487.
Gersternkorn, H. (1955). *Z. Astrophys.* **36,** 245.
Gersternkorn, H. (1957). *Z. Astrophys.* **42,** 137.
Gersternkorn, H. (1969). *Icarus* **11,** 189–207.
Gibson, E. K., Jr. and Hubbard, N. J. (1972). *In* Lunar Science-III. Lunar Science Inst. Contr. No. 88.
Gilbert, G. K. (1893). *Bull. Phil. Soc. Wash.* **12,** 241–292.
Goldreich, P. (1966). *Revs. Geophys.* **4,** 411–439.
Haggerty, S. E. (1972). *In* Lunar Science-III. Lunar Science Inst. Contr. No. 88
Hanks, T. C. and Anderson, D. L. (1969). *Phys. Earth Planet. Inter.* **2,** 19.
Hartmann, W. K. (1970). *Icarus* **13,** 299.
Hinners, N. W. (1971) *Revs. Geophys. and Space Phys.* **9,** 447–522.
Hubbard, N. J. and Gast, P. W. (1971). *Geochim. Cosmochim. Acta Suppl.* **2,** 999–1020.
Hubbard, N. J., Gast, P. W., and Wiesmann, H. (1970). *Earth Planet. Sci. Lett.* **9,** 181–184.

Hubbard, N. J., Gast, P. W., Meyer, C., Nyquist, L. E., and Shih, C. (1971). *Earth Planet. Sci. Lett.* **13,** 71–75.
Hubbard, N. J., Rhodes, J. M., Ridley, W. I., Bansal, B. M., and Wiesmann, H. (1972). Chemical compositions of some Apollo 15 soil samples. (Submitted to *Science, N.Y.*)
Jakes, P., Warner, J., Ridley, W. I., Reid, A. M., Harmon, R. S., Brett, R., and Brown, R. W. (1971). *Earth Planet. Sci. Lett.* **13,** 257–271.
James, O. B., and Jackson, E. D. (1970). *J. Geophys. Res.* **75,** 5793.
Jeffreys, H. (1952) "The Earth". Cambridge Univ. Press, London.
Kaula, W. M. (1969). *Phys. Earth Planet. Interiors* **2,** 123.
Kaula, W. M. (1970). *Science, N.Y.* **166,** 1581–1588.
Kaula, W. M. (1971a). *Revs. Geophys. Space Phys.* **9,** 217–238.
Kaula, W. M. (1971b). *Amer. Geophys. Union Trans.* **52,** IUGG., 1–4.
Keil, K., Bunch, T. E., and Prinz, M. (1971). *Geochim. Cosmochim. Acta Suppl.* **2,** 561–599.
Kopal, Z. (1972). *The Moon* **4,** 28–34.
Langseth, M. G. J., Clark, S. P., Jr., Chute, J. L., Jr., Keihm, S. J., and Wechsler, A. E. (1972). *In* Apollo 15 Preliminary Science Rept., NASA SP-289.
Latham, G. V., Ewing, M. Press, F., Sutton, G., Dorman, J., Nakamura, Y., Toksöz, N., Lammlein, D., and Duennebier, F. (1972). *In* Apollo 15 Preliminary Science Rept., NASA SP-289.
Levin, B. J. (1962). *In* The Moon (Z. Kopal and Z. K. Mikhailov, ed.). Academic Press, New York and London.
Levinson, A. A. and Taylor, S. R. (1971). "Moon Rocks and Minerals". Pergamon, New York.
Lowman, P. D., Jr. (1972). *J. Geol.* **80,** 125–166.
Lunar Sample Preliminary Examination Team (1971). *Science, N.Y.* **173,** 681–693.
Lunar Sample Preliminary Examination Team (1972). *Science, N.Y.* **175,** 363–375.
Lunatic Asylum (1970). *Earth Planet. Sci. Lett.* **9,** 137–163.
MacDonald, G. J. F. (1959). *J. Geophys. Res.* **64,** 1967.
MacDonald, G. J. F. (1964). *Revs. Geophys.* **2,** 407.
MacDonald, G. J. F. and Urey, H. C. (1971). *In* "Physics and Astronomy of the Moon," 2nd ed. Academic Press, New York and London.
Mason, B. and Melson, W. G. (1970). "The Lunar Rocks." Interscience Publishers, New York.
McConnell, R. J., Jr. and Gast P. W. (1972). *The Moon* **5,** 41–51.
McConnell, R. J. R., McLaine, L. A., Lee, D. W., Aronsen, J. R., and Allen, R. V. (1967). *Rev. Geophys.* **5,** 121.
Metzger, A. E., Trombka, J. I., Peterson, L. E., Reedy, R. C., and Arnold, J. R. (1972). *In* Lunar Science-III. Lunar Science Inst. Contr. No. 88.
Meyer, C., Jr. (1972). *In* Lunar Science-III. Lunar Science Inst. Contr. No. 88.
Meyer, C., Jr., Brett, R., Hubbard, N. J., Morrison, D. A., McKay, D. S., Aitken, F. K., Takeda, H., and Schonfeld, E. (1971). *Geochim. Cosmochim. Acta Suppl.* **2,** 393–411.
Muller, P. M. and Sjogren, W. L. (1968). *Science, N.Y.* **161,** 680–684.
Murthy, V. R., Evenson, N., and Hall, H. T. (1971). *Nature, Lond.* **234,** 267–290.
Mutch, T. A. (1970). "Geology of the Moon." Princeton Univ. Press, Princeton, N.J.
Nyquist, L. E., Hubbard, N. J., and Gast, P. W. (1972). *In* Lunar Science-III. Lunar Science Inst. Contr. No. 88.

O'Keefe, J. A. (1969). *J. Geophys. Res.* **74,** 2758–2767.

O'Kelley, G. D., Eldridge, J. S., Schonfeld, E., and Northcutt, K. J. (1972). *Geochim. Cosmochim. Acta Suppl.* **3,** 1659–1670.

Öpik, E. J. (1961). *Astron. J.* **66,** 60.

Papanastassiou, D. A., Wasserburg, G. J., and Burnett, D. S. (1970). *Earth Planet. Sci. Lett.* **8,** 1.

Papanastassiou, D. A. and Wasserburg, G. J. (1971). *Earth Planet. Sci. Lett.* **11,** 37–62.

Pearce, G. W., Strangway, D. W., and Gose, W. A. (1972). *Geochim. Cosmochim. Acta Suppl.* **3,** 2449–2464.

Pearce, G. W., Strangway, D. W., and Larson, E. E. (1971). *Geochim. Cosmochim. Acta Suppl.* **2,** 2451–2460.

Reid, A. M., Ridley, W. I., Harmon, R. S., Brett, R., Jakes, P., and Brown, R. W. (1972). *In* Lunar Science-III. Lunar Science Inst. Contr. No. 88.

Reid, A. M., Ridley, W. I., Harmon, R. S., Warner, J., Brett, R., Jakes, P. and Brown, R. W. (1972). *Geochim. Cosmochim. Acta* **30,** 903–912.

Ridley, W. I., Reid, A. M., Warner, J. L. and Brown, R. W. (1972). (Submitted to *Science*).

Ringwood, A. E. (1970a). *J. Geophys. Res.* **75,** 6453–6479.

Ringwood, A. E. (1970b). *Earth Planet. Sci. Lett.* **8,** 131–140.

Ringwood, A. E. (1972). *In* Lunar Science-III. Lunar Science Inst. Contr. No. 88.

Ringwood, A. E. and Essene, E. (1970). *Geochim. Cosmochim. Acta Suppl.* **1,** 769–800.

Roedder, E. and Weiblen, P. W. (1971). *Geochim. Cosmochim. Acta Suppl.* **2,** 507–528.

Runcorn, S. K. (1972). *In* Lunar Science-III. Lunar Science Inst. Contr. No. 88.

Runcorn, S. K., Collinson, D. W., O'Reilly, W., Battey, M. H., Stephenson, A., Jones, J. M., Manson, A. J., and Readman, P. W. (1970). Proc. Apollo 11 Lunar Sci. Conf. Geochim. Cosmochim. Acta Suppl. 1, 2369–2387.

Ruskol, E. L. (1960). *Soviet Astron,* **AJ4,** 657–668.

Ruskol, E. L. (1963). *Soviet Astron.* **AJ7,** 221–227.

Schmitt, H. H., Trask, N. J., and Shoemaker, E. M. (1967). Map 1–515. U.S. Geol. Survey, Wash., D.C.

Schonfeld, E. (1972). *In* Lunar Science-III. Lunar Science Inst. Contr. No. 88.

Singer, S. F. (1968). *Geophys. J.* **15,** 205–226.

Singer, S. F. (1970). *Trans. Am. Geophys. Union* **51,** 637–641.

Sjogren, W. L., Muller, P. M., and Wollenhaupt, W. R. (1972). *The Moon* **4,** 411–418.

Smith, J. V., Anderson, A. T., Newton, R. C., Olsen, E. J., and Wyllie, P. J. (1970). *J. Geol.* **78,** 381–405.

Soderblom, L. A. and Lebrofsky, L. A. (1972). *J. Geophys. Res.* **77,** 279–296.

Sonnett, C. P., Schubert, G., Smith, B. F., Schwartz, K., and Colburn, D. S. (1971). *Geochim. Cosmochim. Acta Suppl.* **2,** 2415–2431.

Strangway, D. W. (1972). *Trans. Am. Geophys. Union* **53,** 431.

Strangway, D. W., Pearce, G. W., Gose, W. A., and Timme, D. W. (1971). *Earth Planet. Sci. Lett.* **13,** 43–52.

Toksöz, M. N., Press, F., Anderson, K., Dainty, A., Latham, G., Ewing, M., Dorman, J., Lammlein, D., Sutton, G., Duennebier, F., and Nakamura, Y. (1972). *The Moon* **4,** 490–504.

Toksöz, M. N., Solomon, S. C., Minear, J. W., and Johnston, D. H. (1972). *The Moon* **4,** 190–213.

Tozer, D. C. (1972). *The Moon* **5,**
Turner, G. (1972). *Earth Planet. Sci. Lett.* **14,** 169–175.
Urey, H. C. (1952). "The Planets, Their Origin and Development." Yale Univ. Press, New Haven, Conn.
Urey, H. C. (1968). *Science, N.Y.* **162,** 1408–1410.
Urey, H. C. and MacDonald, G. J. F. (1971). *In* "Physics and Astronomy of the Moon." 2nd ed., 213–289. Academic Press, New York.
Vinogradov, A. P. (1971). *Geochim. Cosmochim. Acta Suppl.* **2,** 1–16.
Warner, J. L. (1971). *Geochim. Cosmochim. Acta Suppl.* **2,** 469–480.
Wasserburg, G. J., Turner, G., Tera, F., Posodek, F. A., Papanastassiou, D. A., and Huneke, J. C. (1972). *In* Lunar Science-III. Lunar Science Institute Contr. No. 88.
Wasson, J. T. and Baedecker, P. A. (1972). *Geochim. Cosmochim. Acta Suppl.* **3,** 1315–1326.
Wellman, R. R. (1970). *Nature Lond.* **225,** 716–717.
Wetherill, G. W. (1968). *Science, N.Y.* **160,** 1256.
Wetherill, G. W. (1971). *Science, N.Y.* **173,** 383–392.
Williams, R. J. (1972). *Earth Planet. Sci. Lett.* **16,** 250–256.
Wise, D. V. (1969). *J. Geophys. Res.* **74,** 6034–6045.
Wise, D. V. and Yates, M. T. (1970). *J. Geophys. Res.* **75,** 261.
Wollenhaupt, W. R. and Sjogren, W. L. (1972). *The Moon* **4,** 337–347.
Wood, J. A. (1970). *J. Geophys. Res.* **75,** 6497.
Wood, J. A. (1972). *Icarus* **16,** 229–240,
Wood, J. A., Dickey, J. S., Marvin, U. B., and Powell, B. N. (1970). *Geochim. Cosmochim. Acta, Suppl.* **1,** 965–988.

Physics of Planetary Interiors

G. H. A. COLE
Physics Department University of Hull, Hull, England

Contents

I. Introduction

For the first time man is now making direct physical contact with planets of the Solar System other than the Earth. Automatic probes have landed on Venus, on Mars, and on the Moon, and an artificial orbiting satellite (Mariner 9) has sent back continuous data of the Martian surface and environment over the past year. The series of manned visits to the Moon (Apollo series) have been concluded, and such enormous technical achievements are in danger of becoming commonplace in the popular eye. Close approaches to other planets are planned and the next decade can be expected to be full of excitement from the view point of planetary studies. Although the photographic camera has provided the most immediate general excitement, data collected by other instruments by remote control (such as magnetic field strengths, temperatures, charged particle counts, and so on) will probably have the greatest effect on our ideas in the long run. The recent observations have been possible because of phenomenal technical advances in the two fields of rocketry and of the miniaturization of apparatus, but their especial usefulness lies in the fact that our understanding of the physics of the planets is already well advanced separately on the basis of general principles of physics.

Regarding the Solar System as a single entity, modern studies of the interior of planetary bodies can be claimed with some fairness to have begun in 1897 when Wiechert drew certain broad conclusions about the interior structure of the Earth. The argument was simple (as all good arguments must be) and can be summarized as follows. The total radius, R_p, the total mass, M_p, and the total mean moment of inertia, I_p, of the Earth are all supposed known (see Table 1). The total mean density $\hat{\rho}_p$ is found to be very nearly 5.5 gm cm^{-3} (in all that follows we shall use the same units for the density and need not quote them again) while the mean density of the surface material, ρ_s, is only about half this value. For a body the size of the Earth this large increase in density between the outside and the centre cannot be accounted for by compression alone: a compositional change must be involved as well because the Earth cannot have a homogeneous composition. This conclusion is probed further in terms of a model.

The simplest model (and that studied by Wiechert, 1897) represents the Earth as an isothermal sphere (but see Section V) composed of two regions of different density, the interface between them being a spherical surface of density discontinuity. Suppose the central region, of radius $R_c = xR_p$ and called the core, has a constant density ρ_c while the outside shell, called the mantle, has a constant density ρ_m. We seek those values of ρ_c, ρ_m, and x which provide the observed values of M_p and I_p for given R_p, i.e. of ρ_p and I_p. There can, in fact, be no unique association among these variables because we are attempting to assign values to three variables on the knowledge of two observed variables. The observed data could be fitted by a wide range of mantle and core densities and the particular model appropriate to the Earth can be isolated only on the basis of supplementary information.

Wiechert made the problem tractable by assigning $\rho_m = 3.20$, assuming the mantle to have the density of non-ferrous meteoric material. Accepting from observation the values $\hat{\rho}_p = 5.58$ and $\alpha_p = I_p/M_pR_p^2 = 0.3335$ (the numerical value of the coefficient of inertia is obtained from the precession of the equinoxes: see e.g. Jeffreys, 1962, p. 145; Cook, 1972), we must assign the values $\rho_c = 8.206$ and $x = 0.779$. This value of ρ_c is closely similar to that for meteoric iron at zero pressure and room temperature, but the predicted value of x must also be considered. A decisive check comes from seismology (see Section III) which supports the existence of a deep discontinuity of density but at $x = 0.545$. Jeffreys (1962, p. 154) used this value of x to invert the problem and determine ρ_m and ρ_c: using the improved value $\hat{\rho}_p = 5.53$ and retaining $\alpha_p = 0.3335$, he found that $\rho_m = 4.22$ and $\rho_c = 12.33$. These values treated as mean values are not incompatible with modern data (see Table 8 and Section IX), the difference between the values of ρ_c derived by Jeffreys and by Wiechert being a measure of the effect on the central pressure. Jeffreys pointed out that Wiechert obtained the wrong values for R_c because of the neglect of the compressibilities of the materials of the mantle and core. The seismic data contain this implicitly but a model can include it only if the effects of pressure are accounted for. A detailed commentary of the Earth must be based on a more complicated model.

Although Wiechert's model proved inadequate as a basis for the detailed description of the Earth it is worth recounting because it shows the essential features of present day planetary studies, perhaps as much by the features it ignores as by those that it does not. First and foremost, the internal pressure is a controlling factor. Extending the validity of Wiechert's model for the moment to the study of any planet, it appears that a knowledge of the total mass, total radius and total moment of inertia does not by itself form a sufficient basis for elucidating the interior structure. Other variables must be accounted for. The compressibility of the material is important because, at the simplest level, the stable structure of the planet will result from a balance between the inwardly directed gravitational force and the outwardly directed resistance of the material to compression. Rotation effects (e.g. the action of the centrifugal force) can be ignored to good approximation except in those cases where the compression forces are near the point of inadequacy in relation to the gravitational force (for the Solar System rotation effects are relevant for Saturn). The local pressure in the planet will provide forces which exceed the strength of the material at a relatively shallow depth (beyond about 30 km for the Earth) so that for the major volume of the planet the material will behave as a plastic fluid rather than a rigid solid. This circumstance allows important simplifying assumptions (associated with the condition of hydrostatic equilibrium) to be introduced into the theory, and minimizes the importance of the precise mechanical properties of the constituent material. But such simplifications are bought at the expense of difficulties of detail if the finer structural features of the planet are later of interest. There are two features that offer particular difficulty.

One difficulty is that of gaining appropriate information of the behaviour of minerals under pressure. Laboratory studies can be of great use here providing semi-empirical equations of state of at least limited validity. Of particular relevance for the general theory are those properties common to a range of minerals under planetary conditions although the description of individual planets must involve the properties of those minerals actually present. The elucidation of the mineral content of planets has proved very difficult, and speculation still remains even for the Earth. Laboratory conditions cannot yet include the extreme pressures to be encountered in the giant planets Jupiter and Saturn and it is necessary to appeal to theory for help. Compressibility is in the end a macroscopic manifestation of the strong repulsion between atoms when their electron shells begin to overlap and statistical theories of condensed matter can be invoked to provide theoretical equations of state. The structural complication of the actual materials has so far precluded such fundamental studies from progressing very far. The application of thermodynamic arguments has been more successful in providing formulae of practical utility. These have the additional interest of involving certain properties of materials under general conditions (e.g. those of the Jovian planets) not as yet familiar on the Earth even in the laboratory.

Wiechert's arguments do not involve the temperature. Apparently the effect of temperature is secondary, the planetary structure being essentially isothermal. A body in which effects of the pressure dominate those of the temperature is said to be cold, even though the internal temperature may be several thousand degrees. In this respect, among others, a planet is the opposite extreme of a star, for in a wide class of stars the temperature is a controlling factor (see e.g. Tayler, 1968), such a star being a hot body. There are exceptions, of which a white dwarf star is one, being composed of highly degenerate matter with very high density (typically $\sim 10^6$), a surface temperature of a few thousand degrees, internal conditions controlled principally by the pressure and so showing the characteristics of a cold body. The physics of extreme conditions often provides fortuitous simplifications. For example, if a mass of electrically conducting gas is large enough its effective electrical conductivity is infinitely high even though its local value is not (see e.g. Cowling, 1956); again, the properties of matter at very high temperatures are relatively simple and this fact has allowed stellar studies to progress a long way on the basis of atomic data measured in the laboratory (see e.g. Tayler, 1968). The study of planets faces the difficulty that the internal pressure is either everywhere too low for extreme conditions to apply, or else is extreme only over an insignificant central volume. There are very few quantitative simplifications that can be made.

Of all the members of the Solar System, the Earth has been studied by far the most carefully and many details of the terrestrial interior are known [see Jeffreys' book *The Earth* (1962); also the article by Bullard (1954)]. This has come about over a period of rather more than 200 years through the study of the propagation of seismic waves (see e.g. Bullen, 1963; Jeffreys, 1962, Chapters II and III). One use of this work has been as an aid for the determination of the dependence with depth of such variables as density, elastic wave speeds, Poisson's ratio, the bulk and rigidity moduli, and the pressure. The temperature distribution cannot be found directly this way, but it could presumably be inferred if the precise material composition of the Earth were known together with the dependence of the properties of this material on pressure and temperature. But such a programme would require, for its success, a degree of accuracy quite beyond present possibilities. Although the internal temperature of the Earth is imperfectly known there is every reason to attempt to improve our understanding of it. For one thing it would be possible, if the temperature distribution were known, to obtain information about the distribution of radioactive materials within the Earth. The only direct information for the thermal features of the Earth is the heat flow through the surface together with the temperature gradient in deep mines; presumably this information will, eventually, also be known for at least some of the other planets. The present temperature distribution within the planets will have a direct bearing on the early history of the Solar System but such a cosmogonical study is outside the scope of the present article: the interested reader is referred elsewhere for details (for a recent review see Woolfson, 1969, where there are many references).

Observational data for later use in the paper are collected in Section II. Seismic studies are considered in Section III, while equations of state for planetary material are collected in Section IV. The relation of planetary shapes to theoretical figures of hydrostatic equilibrium is considered in Section V and some consequences of this equilibrium are explored in Section VI. The arguments are applied to the construction of model planets in Section VII. Certain relations between variables are the topic of Section VIII in preparation for the numerical studies of Section IX. The article ends (Section X) with a short discussion.

II. Some Observational Data

The material comprising the Solar System falls naturally into two groups, viz. the planets with attendant satellites, and other bodies such as meteors, comets, asteroids and debris of various kinds. We are concerned here with the planetary bodies alone, which themselves fall into two distinct groups. The first group, called the terrestrial planets, lies within 1.6 A.U.† from the Sun. This group includes the planets Mercury, Venus, Earth, and Mars; there are reasons for treating the Moon as a planetary body, the Earth–Moon system forming a double planet. Apart from a small mass (less than 10^{28} gm) and small radius (less than 10^9 cm), all the planets of this group have mean densities in excess of 3, coefficients of inertia in excess of 0.30, and periods of rotation in excess of 23 h (data are collected in Table 1). The second group, consisting of Jupiter, Saturn, Uranus, and Neptune and called the Jovian or major planets, have properties which contrast with the first group. Thus, all the Jovian planets lie beyond 1.6 A.U. from the Sun, all are massive (with masses in excess of 10^{28} gm), all are large (with radii in excess of 10^9 cm), but all have low densities (less than 2.5). Each member has a rotation period less than 13 h. The values of the coefficients of inertia lie between 0.2 and 0.3, which is less than for the terrestrial planets: for another comparison, a typical value for a star would be 0.05. There are only three naturally occurring satellites associated with the terrestrial planets (i.e. besides the Moon, Phobos, and Deimos are with Mars but the latter two are nearer gigantic boulders than planetary bodies) whereas the Jovian planets have twenty-four satellites and Saturn has its curious triple ring structure. Whereas the terrestrial planets are spherical to a high degree, with oblateness (see Table 2) less than 1/190, the Jovian planets show a marked flattening with an oblateness not less than 1/40. Both groups of planets have solar orbits in the same range of inclination to the ecliptic, the largest angle being 7° for Mercury. The

† A.U. stands for Astronomical Unit: 1 A.U. = 92.9×10^6 miles.

Table 1

Various parameters describing the main planetary members of the Solar System. The second and third columns include data for the total mass and total radius which are used to construct the fourth column for the total mean density. The next column lists values of the mean coefficient of inertia α_p defined by $I_p = \alpha_p M_p R_p^2$, where I_p is the mean moment of inertia. The bracketed values for Mercury and Venus are the result of theoretical calculations and have no empirical basis at the present time. Next, values of the mean radius of solar orbit, L_m, are listed in astronomical units: the rotation period, t_m, in hours is listed, and in the last column we list measured bolometric mean temperatures.

Planet	Mass, M_p, $\times 10^{27}$ gm	Radius, R_p $\times 10^8$ cm	$\hat{\rho}_p$ gm . cm^{-3}	α_p	L_m A.U.	t_m hrs	T_m°C
Mercury	0.324	2.434	5.364	(0.337)	0.387	1440	410
Venus	4.863	6.065	5.227	(0.341)	0.729		430
Earth	5.975	6.371	5.516	0.3335	1	23.933	15
Mars	0.642	3.395	3.942	0.377	1.490	24.624	−30
Moon	0.0735	1.738	3.342	0.398–0.400	1	648	−120
Jupiter	1897.1	69.75	1.335	0.25	5.04	9.9	−150
Saturn	567.7	58.17	0.688	0.22	9.580	10.5	−150
Uranus	86.70	23.75	1.545	0.23	19.21	10.7	−160
Neptune	10.52	22.24	2.283	0.29	29.99	12.8	−170

planet Pluto is left out of these considerations because its orbit is strange and the parameters such as mass and radius are imperfectly known. What indications there are, suggest it to be more like the terrestrial planets than the Jovian, but it is best to disregard it now (but see Cook, 1972, and Seidelmann *et al.*, 1971).

Data are hardly ever varied enough for the theoretician's liking and a

Table 2

Data for the maximum surface density, ρ_s, and oblateness, η. The density ρ_s is derived from equation (8.4) of the text (assuming the planet to have a solid surface), the bracketed values for Mercury and Venus being based on the theoretical values of α_p listed in Table 1. The variable χ is defined by equation (45) of the text.

Planet	ρ_s gm . cm^{-3}	χ	η
Mercury	(4.53)	—	—
Venus	(4.46)	—	—
Earth	4.48	0.00345	1/298
Mars	4.00	0.0043	1/192
Moon	3.325	—	—
Jupiter	0.57	0.0843	1/15
Saturn	0.20	0.1421	1/9.5
Uranus	0.85	0.0546	1/14
Neptune	1.78	0.0194	1/40

wider range of data will accrue in the future from measurements using space vehicles. Existing data will also be improved in accuracy; possibly the least well known of the properties mentioned so far is the moment of inertia. That for the Earth is accurate to about 1 part in 3000, but for Moon and Mars the corresponding accuracy is near 1 part in 300. Nothing is known directly about the coefficients of inertia for Mercury and Venus (these are planets without natural satellites), and the values for the Jovian planets are probably not more accurate than 3% at best.

Planetary bodies are a source of heat. Measurements of heat flux through the surface for the Earth vary with location and are difficult to determine but a mean heat flow through the terrestrial surface of 1.5×10^{-6} cal cm^{-2} sec is probably not grossly in error. The scant information so far available for the Moon, on the basis of recent Apollo measurements, suggest a value there of about one half the terrestrial value—a surprisingly high value, relatively speaking. The mean heat flows for the remaining planets are not known although mean radiation temperatures have been deduced as listed in Table 1. Of course, these temperatures do not necessarily mirror the heat flow through the surface, e.g. the high temperature for Venus is very likely to be an atmospheric green-house effect rather than that of a simple heat flow from inside.

Magnetism is found in various forms throughout the Universe and the magnetism of the Earth is well known (for an extensive review see Chapman and Bartels, 1951; a recent review is Rikitake, 1971; see also Bullard, 1954). The terrestrial field has an essentially dipolar form with a superposed minor non-dipolar component. The field is subject to variation in time, the effect being a slow steady change between periods of collapse and complete reversal. The magnitude of the terrestrial field at the equator is of the order one half Gauss ($\frac{1}{2}$ G). Jupiter alone among the other planets that have been studied so far shows evidence of possessing a magnetic field. The evidence is indirect, involving the non-thermal decimetre and decametre radiation emitted by the Jovian ionosphere, and the strength and configuration of the field are essentially unknown. Assuming the field to be dipolar, the strength of the emission depends primarily upon both the field strength and the density of the plasma in the atmosphere of the planet. A knowledge of the one can lead to information about the other, but so far little is known about either quantity. Values of field strength between 1 and 100 G at a distance of one radius about Jupiter's surface have been envisaged but only speculatively, although the lower value would seem the more likely on energy considerations. There are indications that the direction of the dipole is opposite to that of the Earth, and that it makes a small angle (perhaps about 9°) with the rotation axis. Space probe measurements, involving magnetometer measurements down to 10^{-6} G, have failed to provide any direct evidence for the existence of magnetic fields with sources within the Moon, Mars or Venus. There is the curious result of the Apollo missions that local weak lunar fields with strengths of the order 10^{-3} G may have some permanence, and paleomagnetic studies of lunar material show evidence for the existence of a general lunar field in the distant past with a strength of a few percent of that of the present terrestrial field. Certain meteorites show a weak magnetism that cannot, apparently, be linked directly to the effects of terrestrial magnetism. The remaining planets and satellites have not been studied directly in these terms but what evidence there is does not suggest that these bodies are the source of any significant magnetic fields. It would seem safe to conclude that magnetism is unlikely to play any significant role in the dynamical behaviour of the planets, although the account of the magnetic phenomena can provide at least indirect information about the dynamical processes occurring inside.

III. Seismic Studies

Seismology has proved an invaluable empirical method for exploring the deep interior of the Earth treated as an elastic deformable body (Bullard, 1954; Jeffreys, 1962; Bullen, 1963). These studies have the important significance from the present point of view of supporting to a surprisingly high degree of approximation, the representation of the main bulk of the Earth as a simple elastic medium, at least under the stimulus of seismic waves. The application of the simple theory could be an obvious first approximation for other planetary bodies as well, corrections to account for more complex behaviour being applied when necessary as further approximations.

The classical theory of simple elasticity describes the local elastic behaviour, for prescribed thermodynamic conditions, in terms of two parameters, conveniently the adiabatic bulk modulus K_s and the rigidity μ. For a homogeneous isotropic infinite medium showing perfect elasticity two non-dispersive and non-dissipative wave modes are found, a longitudinal P-wave field and a transverse S-wave field, with local speeds V_p and V_s respectively. If ρ is the local density, theory gives the formulae

$$V_p^2 = \frac{K_s + \frac{4\mu}{3}}{\rho}, \qquad V_s^2 = \frac{\mu}{\rho}. \tag{1}$$

It is clear that $V_p > V_s$ and that $V_s = 0$ when the rigidity is zero. Great care must be taken not to make too ready an interpretation of such statements in terms of immediately available materials. Thus, a fluid is characterized by no rigidity on the ordinary hydrodynamical time scale, but still shows rigidity for wave frequencies high in comparison with the inverse of the relaxation time of the fluid. This behaviour is well shown by cold pitch which is rigid to impulsive impact and will transmit transverse waves except of the lowest frequency: but it will show all the properties of normal flow (except that the time scale is slowed down) if subjected to even a small stress for a sufficient length of time (which may extend into weeks). Again, glass is a very viscous liquid even at room temperature. Poisson's ratio σ follows uniquely from a knowledge of V_p and V_s. For, theory provides the expression

$$\sigma = \frac{3K_s - 2\mu}{2(3K_s + \mu)} = \frac{v^2 - 2}{2(v^2 - 1)}, \tag{2}$$

where in the last expression $v = V_p/V_s$.

The wave propagation is affected by inhomogeneities and the trajectories will be affected. Of especial importance is a boundary of density discontinuity, where the elastic waves are partly reflected and partly refracted according to the requirements of the conservation of wave energy and momentum. A pure wave (either P or S) will have its energy partitioned by the boundary into both P and S forms except for normal incidence or for those angles for the transverse wave for which its vibrational direction is parallel to the surface. One consequence is the appearance of S waves in a region of non-zero rigidity even though the wave disturbance may have entered as a pure P-wave mode having passed through a region of zero rigidity, where S-waves cannot penetrate. For a general body with material inhomogeneities, including chemical variations, the situation is particularly complicated, the wave modes becoming dispersive; the full pattern is a mixture of P and S waves and combination waves (such as SP, SPS, PSP, etc) which requires great skill to disentangle into component members.

The application of simple theory can be applied to good approximation when the material is isotropic over a volume of characteristic length large in comparison with that of the seismic waves. This can occur when the local pressure exceeds the strength of the material and is generally the case for pressures in excess of about 3×10^9 dyne cm^{-2} which for the Earth are achieved for depths greater than some 30 km. The material below this depth is the mantle while above it is the crust. The same general conclusions can apply to each planet with a solid surface. The details of the crustal region will presumably differ from one planet to another. For the Earth, the crust has a layered structure which carries surface waves whose properties were investigated particularly by Rayleigh and Love. Such a layered structure cannot be expected to be reproduced in other planets (see especially Section IX,A,2).

The emphasis to this point has been on the propagation of elastic energy through the Earth by local wave modes. The frequencies involved here are of the order of a minute or two, but a wider range of frequencies are also present associated with the free vibrations of the Earth as a whole (Bullen, 1963,

p. 250). This wider range of frequencies cannot be adequately observed by using conventional seismometers but instead requires more refined techniques for measuring small ground movements. These have been developed during the last ten years but their potentialities still remain to be full exploited. The theory of free oscillations had, quite separately, been developed extensively long before this, starting with the studies by Poisson about 1829 of the vibrations of a perfectly elastic solid sphere. Various modes of vibration have been recognized including radial, spheroidal, and torsional oscillations, particularly for a spherically symmetric, homogeneous body. Deviations from spherical shape and the effects of inhomogeneities are included as perturbation effects. The fundamental and lower harmonics have a period of an hour or so; the short period P, S and surface seismic disturbances, of strictly local association, are then harmonics of very high order (perhaps in excess of 140). Further effects of very long period arise from forced tidal oscillations due to the gravitational attraction of the Moon and Sun. A complicated range of frequencies of oscillation are found for the Earth the details of which are determined by the structure of the Earth including the core. The interpretation of these measurements provides a powerful tool for the detailed study of the Earth (Bullen, 1954; Dziewonski, 1971), including dissipative processes (Sato and Espinosa, 1967a and b).

These various methods have recently begun to be applied to the Moon by the study of seismic waves and presumably will soon be applied to Mars as well. Application to Venus could reasonably be anticipated in the future and a wide knowledge of the interior conditions of the terrestrial planets can be anticipated ultimately on this basis.

IV. Equations of State

The local elastic behaviour of the planetary material is dictated by the stressed condition due to the weight of material above. The strength of terrestrial mantle minerals can be assumed to be of the order of a few times 10^9 dynes cm^{-2} and the pressure is in excess of this at a relatively shallow depth below the surface. Below this depth the material has effectively no permanent rigid strength and behaves as a very viscous fluid, with a relaxation time running into millions of years. Above it, in the crust, the material strength is crucial and the material shows the characteristic features of a rigid body. We shall exclude the crust for the present.

The specification of the stress conditions in the mantle is conveniently made in terms of an equation of state relating the pressure to the density and perhaps also the temperature (see Table 3). In general terms such an equation will be complicated although ideally it should be deducible from the principles of solid state physics. In practice it is both convenient and sufficient to

Table 3

Equation of state data for three materials of interest for terrestrial type planets. T_s refers to the adiabatic temperature with depth in the Earth, with gradient calculated from equation (A4.3).

	Material						
	Rock		Corundum		Periclase		
Pressure $\times 10^9$ dyne cm^{-2}	ρ gm cm^{-3}	$\beta \times 10^6$ deg^{-1} K	ρ gm cm^{-3}	$\beta \times 10^6$ deg^{-1} K	ρ gm cm^{-3}	$\beta \times 10^6$ deg^{-1} K	T_s °K
0	3.16	18.4	3.98	20.5	3.58	26.3	300
0.050	3.21	24.9	3.92	28.6	3.55	30.3	1532
0.100	3.31	22.7	3.94	29.9	3.61	29.3	1852
0.150	3.64	20.0	3.98	29.6	3.68	27.4	2077
0.200	3.80	18.3	4.04	28.4	3.75	25.5	2256
0.250	4.20	16.2	4.09	26.6	3.83	23.8	2407
0.300	4.25	15.1	4.15	24.6	3.90	22.3	2539
0.400	4.35	13.9	4.26	21.2	4.04	19.9	2764
0.500	4.46	12.8	4.36	18.6	4.17	17.8	2954
0.800	4.75	10.4	4.60	16.9	4.52	13.2	3404
1.00	4.91	9.36	4.76	16.5	4.70	11.3	3644
2.00	5.69	7.55	5.62	14.4	5.52	9.64	4513
3.00	6.45	6.54	6.49	11.8	6.42	8.62	5123
4.00	7.17	5.48	7.29	9.55	7.27	7.05	5609

invoke empirical relationships. The form to be selected for such relationships will depend upon the time scale involved. The equilibrium configuration, referring to times in excess of the relaxation time of the material, a static approximation is adequate, the internal stresses being represented by an hydrostatic pressure. For times smaller than the relaxation time, the planet must be treated as a dynamical structure and the stress must be accepted as the plastic flow of very viscous material (Orowan, 1965; Ryabinin, *et al.* 1971).

A. STATIC APPROXIMATION

It is convenient to consider the material conditions within the terrestrial planets separately from those within the Jovian planets. This is partly because the pressure ranges are different in the two groups (see Table 7) and partly because the relative material abundances differ. But in either case we are dealing with a mixture of elements and not a single constituent.

1. *Low and Medium Pressures*

The pressures here are not in excess of 10^{12} dynes cm^{-2} and the temperatures do not exceed a few thousand degrees. The isothermal bulk modulus is defined by

$$K_T(p) = \rho \left.\frac{\partial p}{\partial \rho}\right|_T; \tag{3}$$

alternatively, $K(p)$ can be expanded in powers of the pressure about the value for zero pressure K_0 in the form

$$K_s(p) = K_0 + \sum_j B_j p^j, \tag{4}$$

$$B_j = \frac{1}{j!} \frac{\partial^j K_0}{\partial p^j}. \tag{5}$$

The expressions (3) and (4) can be combined to provide empirical relationships between the pressure p and density ρ. The simplest case is when the expansion (4) is terminated at the linear term, i.e. $B_1 = B$ and $B_j = 0$ for $j \geqslant 2$. Then integration of equation (3) provides the isothermal equation of state (Murnaghan, 1944, 1951).

$$p = \frac{K_0}{B}[y^B - 1], \qquad y = \frac{\rho}{\rho_0}, \tag{6}$$

where ρ_0 is the density at zero pressure. This equation of state has been widely used for planetary problems with the index B appropriately chosen. For laboratory experiments using terrestrial mantle constituents many of the data are well represented by $B \sim 4$ (see Table 4); for pure iron under conditions appropriate to the terrestrial core it is better to take $B = 3.3$. The values of K_0 and ρ_0 will generally be changed to account for phase change due to pressure, but the value of B could remain unchanged to good approximation, although experimental evidence is not easy to obtain on this point. It should be noticed that K_0 and ρ_0 need have only a formal meaning since the phase to which they refer need not actually exist at zero pressure.

While equation (6) can be useful for the lower pressures, further terms in the expransion (4) must be included as the pressure increases. Thus, Bullen (1963) has included also the quadratic term in connection with the lower mantle region where the pressure is of the order 10^{12} dyne cm^{-2}. Although the coefficients B_1 and B_2 can be chosen to provide the best fit for independent data (e.g. seismic travel times), the coefficients B_1 and B_2 should more properly be selected on the basis of the dependence of K_0 on the pressure according to equation (5). A complicated p-ρ relationship will now follow from equation (3), dependent upon the pressure and reducing to the form (6) for sufficiently low pressures.There are still only limited laboratory data to tell of the behaviour of matter at high pressures and in consequence further hypotheses are unavoidable. One such hypothesis (due to Bullen, 1946, 1949) and based on empirical terrestrial seismic travel-time data is that, to an acceptable degree of approximation, the bulk modulus is a smoothly varying function of the pressure, for pressures of the order of a million atmospheres. Earth models constructed on this assumption (called by Bullen Models A),

Table 4

A selection of elastic and other data for ten materials of interest for terrestrial type planets. The second column involves data for the adiabatic bulk modulus at zero pressure, $K_s(0)$, while the third column contains data for the density at zero pressure, ρ_0. The next two columns contain data for the dependence of $K_s(0)$ on the pressure and on the temperature. Θ is the Debye temperature, calculated using equation (13) of the text, while σ is the Poisson's ratio and W the atomic weight of the material in grammes.

Material	$K_s(0)$ $\times 10^{12}$ dyne cm^{-2}	ρ_0 gm cm^{-3}	$\left.\frac{\partial K_s(0)}{\partial p}\right\|_T$ $\times 10^9$ dyne cm^{-2} deg^{-1} K	$\left.\frac{\partial K_s(0)}{\partial T}\right\|_p$ $\times 10^9$ dyne cm^{-2} deg^{-1} K	Θ °K	σ	W gm
Corundum (α-Al_2O_3)	2.521	3.97	3.98	−0.14	1029	0.236	20.39
Zincite (ZnO)	1.411	5.62	4.78	−0.13	410	0.357	40.68
Periclase (MgO)	1.624	3.58	4.58	−0.16	946	0.186	20.16
Spinel ($MgO.Al_2O_3$)	2.020	3.62	4.18	−0.13	887	0.260	20.36
Forsterite (Mg_2SiO_4)	0.973	3.02	4.87	−0.11	647	0.284	20.10
Garnet	1.770	4.16	5.43	−0.20	745	0.274	23.79
Lime (CaO)	1.059	3.29	5.23	−0.14	654	0.210	28.04
Quartz (α-SiO_2)	0.377	2.649	6.40	−0.10	572	0.077	20.03
Bromellite (BeO)	2.201	3.000	5.52	−0.12	1274	0.205	12.51
Hermatite (α-Fe_2O_3)	2.066	5.274	4.53	−0.17	660	0.308	31.94

reproduced seismic data well, but the comparison is improved further if in addition it is supposed that $(\partial K/\partial p)$ is also a smoothly varying function (Models B). Of course, such empirical assertions can be proposed in large numbers and a proper discussion must be based on theory, such as that involving the free energy of the material.

One such approach is that of Murnaghan (1951) who recognized that the classical theory of infinitesimal strain is inappropriate for conditions deep within a planet where materials undergo finite deformations. The hydrostatic strain $(-f)$ within an isotropic medium with cubic structural symmetry (such as garnet) is related to the density by

$$y = (1 + 2f)^{3/2}; \tag{7}$$

it will be realized that the variable f is defined to be positive for compression. Suppose the initial state is devoid of strain, and assume the free energy to be expressible as a power series in f. Using the formulae of thermodynamics and terminating the expression for the free energy at the cube power of f the formulae for isothermal compression result (see Appendix 1):

$$p = 3K_0 f(1 + 2f)^{5/2}(1 - 2\xi f), \tag{8}$$

$$K_T = K_0(1 + 2f)^{5/2}[1 + 7f - 2\xi f(2 + 9f)], \tag{9}$$

$$\left.\frac{\mathrm{d}K}{\mathrm{d}p}\right|_T = \frac{12 + 49f - 2\xi(2 + 32f + 81f^2)}{3[1 + 7f - 2\xi f(2 + 9f)]}. \tag{10}$$

The equation (10) can be regarded as defining the function ξ. The equations (8)–(10) neglect all terms involving powers of ξ higher than the first. The theory has been applied to terrestrial problems by Birch (1952, 1961), and for these purposes it is convenient to replace the strain in equation (8) by the density according to equation (7). The result is the Birch-Murnaghan equation of state,

$$p = \tfrac{3}{2}K_0(y^{7/3} - y^{5/3})[1 - \xi(y^{2/3} - 1) + \cdots]. \tag{11}$$

For conditions of small strain equation (10) becomes (in the limit $f \to 0$)

$$B_1 = \left.\frac{\mathrm{d}K_0}{\mathrm{d}p}\right|_T = 4 - \tfrac{4}{3}\xi, \tag{12}$$

while if ξ is zero, $B_1 = 4$ in agreement with experimental data for moderate pressures. It will be realized that the functions K_0 and ξ are each dependent on the temperature.

Equation (11) provides a good representation of cubic compounds such as garnet, periclase, and fluorite, and for metals, the function ξ being virtually zero. This is not so for materials other than of cubic symmetry although equation (11) may still provide a satisfactory description of the state. The Birch-Murnaghan equation has been widely used (but sometimes in modified form) in many recent analyses of planetary interiors (see Section IX). Alternative equations of state have also been constructed, some by considering different definitions of the strain (e.g. see Knopoff and Uffen, 1954), some by including higher terms in the equations (8)–(10). Still others (see Appendix 2) have been obtained from a more correct expression of the free energy, for an aggregation of spatially ordered particles including also anharmonic contributions to the interparticle energy (Thomsen, 1971). In numerical terms, all these equations of state are essentially indistinguishable (see Appendix 1) to the accuracy of measurement appropriate at the present time and we will not consider them further here (see also Thomsen and Anderson, 1969).

The Debye temperature Θ of the material is interesting because it is a measure of the vibrational frequencies for the atomic particles of the material. Specifically

$$\Theta = \frac{h}{k_B}\left[\frac{9N\rho}{4\pi W}\left(\frac{1}{V_p^3} + \frac{2}{V_s^3}\right)^{-1}\right]^{1/3}, \tag{13}$$

where h is the Planck constant, k_B the Boltzmann constant, N is Avegadro's number and W is the atomic weight of the material. The values of C_v can be constructed, once Θ is known, using the theory of specific heats (Joos, 1964).

Anharmonicities of a crystal lattice are described by several parameters of which the adiabatic Grüneisen parameter γ_s is perhaps the most important. Explicitly:

$$\gamma_s = \frac{\beta_p}{C_p}\frac{\partial p}{\partial \rho}\bigg|_s, \tag{14}$$

where β_p is the coefficient of thermal expansion and C_p is the specific heat at constant pressure. Using equation (11), equation (14) becomes

$$\gamma_s = \frac{(5y^{2/3} - 2) - \frac{1}{3}(28y^{4/3} - 30y^{2/3} + 6)\xi}{(5y^{2/3} - 3) - (7y^{2/3} - 3)(y^{2/3} - 1)\xi}. \tag{15a}$$

For the datum limit of zero strain ($y = 1$) equation (15a) becomes

$$\gamma_s(0) = \tfrac{3}{2} - \tfrac{2}{3}\xi. \tag{15b}$$

With the Grüneisen parameter available, thermodynamics (Hoare, 1946) provides a relation between the specific heats at constant pressure and constant volume, viz.

$$C_p = C_v(1 + \beta_p \gamma T). \tag{16}$$

It follows immediately that the adiabatic and isothermal bulk moduli are related by

$$\frac{K_s}{K_T} = \frac{C_p}{C_v} = 1 + \beta_p \gamma T. \tag{17}$$

In seismic studies (see Section III) it is conventional not to distinguish between the adiabatic and isothermal moduli (see data collected in Table 5),

Table 5
Containing data for nine different materials for the ratio of the adiabatic to isothermal bulk moduli, K_s/K_T, the thermal expansion at constant pressure β, the specific heat at constant pressure, C_p, and the adiabatic Grüneissen constant.

Material	K_S/K_T	β_p deg^{-1} K $\times 10^{-6}$	C_p etg gm^{-1} deg^{-1} K $\times 10^{-6}$	γ_s
Corundum	1.006	16.3	7.83	1.20
Zincite	1.003	15.0	4.57	−1.25
Periclase	1.014	31.5	9.25	1.46
Spinel	1.005	16.2	8.03	0.49
Fosterite	1.006	24.0	8.38	0.88
Garnet	1.008	21.6	7.61	1.13
Lime	1.095	28.1	7.64	0.97
Bromellite	1.007	17.7	10.22	0.43
Hermatite	1.019	32.9	6.50	0.64

and consistency would require that no distinction be made between the corresponding specific heats either, to this level of accuracy.

These various formulae must be applied not to a single component material but to a multi-component mix, and it is necessary to take account of this fact. For pressures that are not too high (not exceeding a few times 10^{11} dynes cm^{-2}) it is sufficient to invoke an assumption of additivity of the partial molar *volumes*. If ρ_m is the density of the mixture, ρ_j the density of the *j*th constituent and w_j the mass fraction of the *j*th component then,

$$\frac{1}{\rho_m} = \sum_{j=1}^{n} \frac{w_j}{\rho_j} \tag{18}$$

for an n-component rock mixture. Great care must be exercised in using this expression in practice (see Section IX.A.1). This superposition hypothesis can be expected to hold for certain other quantities such as energy, so that specific heats at constant volume will follow a similar rule to the density.

Where the decomposition expression (18) for the density applies, the coefficient of thermal expansion for the mixture β_m is expressed as

$$\beta_m(p, T) = \rho_m(p, T) \sum_{j=1}^{n} \frac{w_j \beta_j}{\rho_j}. \tag{19}$$

These various formulae, together with others that we do not set down here, have been used for the description of the Earth for some time; the new recent developments are their use in connection with the planets generally.

2. *Extreme Pressures*

The behaviour of materials at extreme pressures requires a different description. Now the electron clouds overlap, and the atoms suffer pressure ionization. The free electrons provide a mechanism for easy electrical conduction and materials that are dielectrics at lower pressures (such as iodine or phosphorus) show the conduction properties of metals if the pressure is high enough. There is no unambiguous evidence that the condition of pressure ionization represents a new thermodynamic state with corresponding discontinuities in the pressure-density and pressure-energy relations. It is now thought unlikely that increases in pressure beyond some 5×10^{11} dyne cm^{-2} are necessarily associated with substantial discontinuous increases of density such as is found at the mantle-core boundary of the Earth. It is, then, unlikely that the terrestrial core is simply a higher pressure modification of mantle material—the assignment of a change of composition from rock to iron at this interface seems almost inescapable. The uncertainties arise in part from the fact that pressures in excess of some 10^{10} dyne cm^{-2} can be achieved in the laboratory only dynamically (using shock wave techniques) whereas our present interest for planets is in hydrostatic pressures. For the higher pressures, therefore, data along a Hugoniot line must be transformed to adiabats using theoretical arguments, data being obtained in an appropriate form for planetary studies only indirectly (Davies and Anderson, 1971; Ahrens *et al.*, 1971). Although it would seem that the progressive increase of pressure ionization is not associated with marked density discontinuities in general, one possible exception is the change of hydrogen from the dense molecular to the metallic phase where a density increase as large as 20% is not impossible (Wildt, 1954).

Once pressure ionization is complete, with the innermost electronic shells ionized, the equation of state of Fermi-Dirac can be applied, although at lower pressures the Birch-Murnaghan equation is possibly a good approximation. If the electrons are treated as classical particles obeying the Fermi-Dirac statistics, then if the pressure is high enough the pressure-density relation at zero temperature is found to be (Gilvarry, 1966)

$$p = D_0 \rho^{5/3} - D_1 \rho^{4/3} \tag{20a}$$

where

$$D_0 = \frac{h^2}{5m}\left(\frac{3}{8\pi}\right)^{2/3}\left(\frac{Z}{A}\right)^{5/3}, \qquad D_1 = \left(\frac{32\pi^3 e^6}{125}\right)^{1/3} A^{4/3} Z^2. \tag{20b}$$

In addition to the usual notation, m is the electron mass, and Z is the atomic number and A is the atomic weight. Further approximations, involving the Schrödinger equation, have been considered (Simcox and March, 1962) but we will not pursue these matters further here.

The bulk modulus which corresponds to the equation of state (20) is

$$K = \tfrac{5}{3} D_0 \rho^{5/3} - \frac{4}{3}\frac{D_1}{D_0}\rho^{4/3} \tag{20c}$$

At very high pressures, $\partial K/\partial p \rightarrow \frac{5}{3}$. These expressions are restricted to zero temperature but are only modified slightly in numerical terms when a finite temperature is included.

The relations (20) will apply for pressures exceeding some 10^{13} dyne cm^{-2} while laboratory static data are available up to some 10^{10} dyne cm^{-2}. Unfortunately, pressures of planetary interest lie in this gap. The problem is a critical one and has been discussed by a number of authors (Wildt, 1954; de Marcus, 1958): it is made more complicated in that several materials are involved in a mix rather than a pure material. It is probably best to assign a representative atomic number to the material (Knopoff and Uffen, 1954), which incidentally need not be integral, and join a Birch-Murnaghan type formula (11) to the form (20) asymptotically at a pressure of about 10^{12} dyne cm^{-2}. This procedure would seem to be unambiguous for materials such as fayalite and forsterite, but SiO_2 itself presents problems that we cannot explore here (McQueen *et al.*, 1967; Cook, 1972).

Temperature effects have been neglected so far because most of the theoretical work has referred to the absolute zero of temperature. It would seem however, that the effect of temperature is not critical (Knopoff and Uffen, 1954).

The behaviour of hydrogen and helium under high pressures is of interest for the Jovian planets (see Section IX.B), and this has been summarized by

De Marcus (1958). These data are indicated in Table 6. They can be represented fairly accurately by a Birch-Murnaghan type formula (Stewart, 1956), and reasonably satisfactorily by a Murnaghan type formula [equation (6)] with $B \sim \frac{8}{3}$. The transition from molecular to metallic hydrogen is likely to occur at a pressure of about 2–3 × 10^{12} dyne cm^{-2} (Wildt, 1954).

Table 6
Pressure-molar volume data for molecular hydrogen and helium, based on calculations of De Marcus (1958).

Pressume dyne cm^{-2}	Molar volume cm^3 per mole H_2	He
0	22.65	—
200	21.0	15.8
400	20.0	14.5
600	19.2	13.5
1000	18.0	12.4
3000	15.1	9.9
6000	13.2	8.5
10000	11.8	7.5
20000	10.1	6.4

There is interest in heavier molecules, and particularly methane and ammonia. Of particular relevance is the formation of metallic ammonia by the pressure ionization of a mixture of ammonia and hydrogen, the transition pressure for which may not be in excess of 10^{11} dyne cm^{-2} (Bernal and Massey, 1954).

B. DYNAMICAL APPROXIMATIONS

Recently the first steps have begun to be taken towards the construction of dynamical planetary models able to describe plastic movements such as mantle convection. The commonly occurring terrestrial minerals show a steady creep under stress but with a specific, lower activation energy U_0. The relevant quantity for fluid flow is the time rate of strain rather than the strain itself, and this can be expected to be proportional to exp $(-U/k_B T)$ where U is the activation energy for movement at temperature T. If δ is the stress, a linear relation between U and δ can be expected at high values of the stress, i.e. $U = U_0 - \mathbf{b}\delta$. For this case, the time rate of strain is proportional to the factor exp $(-U_0/k_B T) \times$ exp $(-\mathbf{b}\delta/k_B T)$. This type of expression will not be expected to hold for low values of the stress, and there it is empirically preferable to replace the exponential dependence on δ by a simple power expression like δ^n with $n \sim 3$. The theory of macroscopic elastic liquids is still under development (see e.g. Andrade, 1947; Lodge, 1964), and although the application of this work to planetary problems is still in the future it promises interesting possibilities (Franck, 1968; Weertman, 1970; Elsasser, 1971a, b). But the study of dynamical behaviour can proceed to some extent on the basis of the properties of more simple fluids, using the well-established principles of Newtonian fluid mechanics. The consequence of the acceptance of the conservation of fluid mass, momentum, and energy is a pattern of convection currents which could extend over appreciable volumes of the interior. The fluid viscosity will inhibit the motion, the condition for convection being that the buoyancy force (proportional to the temperature difference between two layers in the fluid) shall be of sufficient magnitude to overcome the viscous drag. The ratio of the buoyancy to the viscous forces is used to define a dimensionless grouping of variables called the Rayleigh number, Ra, according to

$$\mathrm{Ra} = \frac{\beta g L^4 \rho C}{\nu \lambda}\left(\frac{\Delta T}{L}\right).$$

Here g is the acceleration due to gravity, C the specific heat of the material (whether at constant volume or pressure is irrelevant since the material is essentially incompressible for the present discussion), ν is the kinematic viscosity, λ the thermal conductivity, and ΔT the temperature difference over the characteristic length of the motion, L. Alternatively, the ratio of the buoyancy to the gravitational forces is used to define the Grashoff number G according to

$$G = \frac{\beta g \, \Delta T L^3}{\nu^2}.$$

If it is supposed that the representation of the planetary material as a simple fluid is a sufficient approximation, critical values of Ra and of G concerning the onset of convection are open to calculation (actually within a linear approximation). Such values are then available for the deduction of the properties of the planetary material (e.g. viscosity) at least as to orders of magnitude.

V. Relation to Figures of Equilibrium

The observed figures of the planets (as opposed to bodies of asteroidal size or below) are closely similar to that of a mass of fluid material in steady rotation (Chandrasekhar, 1971). The similarity cannot be pressed beyond a certain accuracy because when viewed in extreme detail, the planetary surface shows local deviations about a hypothetical smooth figure of rotation. But such deviations are sufficiently small to make it natural to consider the rotation figure as a zero order approximation to the true figure, further approximations allowing the actual planetary shape to be approached as closely as appropriate. We consider the planetary shape in these terms now. The problem was stated precisely by Darwin (1899) who enquired of the mean radius and moment of inertia of the equilibrium configuration of a given mass of material of prescribed elastic properties (and particularly of known compressibility) rotating with an assigned angular speed. Preliminary discussion can usefully be based on an isothermal model although the analysis could also involve the temperature distribution and its time evolution.

The equatorial regions of a spinning body will be drawn out by centrifugal forces and the poles will consequently be drawn in: the body shows a flattening described as oblateness which is measured by the difference between the equatorial and polar radii. The result is to induce a value I for the moment of inertia about the axis of rotation different from that, denoted by A, about an axis in the equatorial plane. Oblateness gives rise to a small perturbation in the gravitational field surrounding the body (of the order of 1 part in 1000 for the Earth) the effect of which is to perturb the trajectories of satellites whether natural or artificial (Cook, 1971; Wildt, 1954; De Marcus, 1958). This is the sole effect for a planet in hydrostatic equilibrium, where the surfaces of constant density are also equipotential surfaces for gravitation. The surface elliptic shape of the planet (specified by an ellipticity) can be inferred from observations of the surrounding external gravitational field. Non-hydrostatic effects also arise but their effect on the surrounding field is small; for the Earth, they are smaller than the oblateness effects by a factor of 10^{-3}.

The gravitational field deduced from observations can be associated with a distribution of density within the planet although it is not possible to deduce a unique density distribution from a specified external field. Whereas an assumed density distribution will lead to a unique ellipticity, a measured ellipticity will generally be compatible with a wide range of density distributions. Other variables follow the same lack of reciprocity of viewpoint. Thus information about the change of phase with pressure cannot be deduced from external measurements; a comparable situation applies for the temperature because an assigned distribution of internal heat sources will define the surface outward-flux of heat but the heat flux itself cannot be used to isolate a unique distribution of heat sources.

The expression of the relationship between a specified density distribution and the external gravitational field is achieved by employing the well established theory based on the symmetry described by spherical harmonic functions (see Appendix 3). There are several steps. First the actual planetary figure is replaced by an equivalent ellipsoid, including a measure of the deviation of the true figure locally from the ellipsoid. Next the ellipsoid is replaced by a sphere, of radius R_E, of the same volume as the ellipsoid. In this way the actual figure is reduced to the more amenable symmetry of the sphere. The potential outside the sphere is expressed as an expansion in the spherical harmonic angle functions. The first term in the expansion is that of a point mass at the centre and successive terms introduce corrections due to the actual shape. The coefficient of the second term involves the quantity $J_2 = \dfrac{I - A}{M_p R_E^2}$ where M_p is the total mass of the planet, and so has a simple physical interpretation; the further coefficients cannot be interpreted so directly. The difference $(I - A)$, expressed as the dynamical ellipticity, can be deduced for the Earth from measurements of the rate of precession of the planetary axis; if J_2 is known from measurements, the mean coefficient of inertia $\alpha_E = I/M_p R_E^2$ about the polar axis can be inferred (see also Section IX.B.1).

The theory of the figure of a planet in hydrostatic equilibrium with a prescribed internal density distribution has been developed principally by Darwin (1899), Radau (1885), and de Sitter (1924); De Marcus (1958) has more recently arranged it into a form showing explicit dependence of the equation of state. The method involves an expansion in powers of the centrifugal force and leads immediately to an expression for the coefficient of inertia. Explicitly, we have for a spherical body,

$$\alpha_p = \frac{I_p}{M_p R_p^2} = \frac{8\pi}{3 M_p R_p^2} \int_0^{R_p} \rho(r) r^4 \, dr \tag{21}$$

which is re-expressed, using the theory, into the form at the surface

$$I_p = \frac{2}{3}\left[1 - \frac{2}{5}\left(\frac{5}{2}\frac{\chi}{\varepsilon_s} - 1\right)^{1/2}\right]. \tag{22}$$

Here $\chi = \omega^2 R_p^3 / G M_p$, being the ratio of the centrifugal force to the gravitational force at the equator and ε_s is the surface ellipticity. Equation (22) is only approximate but the reader is referred elsewhere for details (Wildt, 1954; De Marcus, 1958).

The arguments outlined so far have been based strictly on conditions of hydrostatic equilibrium and it is important to assess the usefulness of this restriction in practice. Of course, one test of the usefulness of the formulae is the accuracy with which they predict the magnitudes of quantities that can be measured but this is a somewhat separate issue. Again it might be argued that axial symmetry implies conditions of hydrostatic equilibrium, and the planets appear to be figures of revolution to a high approximation. The pressure inside even the smaller planetary bodies (such as the Moon) will exceed the strength of the constituent material locally beyond a depth of some 30–80 kilometres so that the material can be expected to behave as if plastic for greater depths, the stress differences in the material being a small fraction of the mean pressure. This effect has already been used as a means of distinguishing the crust from the true mantle. These ideas can be checked for the Earth and the Moon using earthquake studies, since earthquakes can be taken as evidence of non-hydrostatic processes locally. As a general rule for the Earth (Bullen, 1965, p. 278), earthquake hypocentres occur most frequently above 30 km, relatively frequently in specific areas down to 100 km and only in isolated cases below this. The deepest hypocentre recorded so far was a little more than 700 km. Data for the Moon are more scant but moonquakes have been recorded at the three sites where measurements have been made (Press, 1971). Although detail is lacking it is clear that seismic activity on the Moon is considerably less than on the Earth. Occurrences are found particularly at monthly intervals corresponding to perogee and apogee of the orbit. Present indications are that the sources are at depths below the lunar surface not in excess of 40 km and are associated with the mascons: this can be taken as indicating that the mascons are not in isostatic equilibrium with the surroundings. Seismic evidence is not apparently in conflict with the assumption that the main lunar volume is in hydrostatic equilibrium to sufficient approximation for our present purposes.

VI. Some Consequences of Hydrostatic Equilibrium

In order to make progress with the application of the laws of physics to the study of planetary interiors we will accept the restriction to conditions of hydrostatic equilibrium. This means that in all that follows we disregard the crust, treating the planet as a symmetry figure of rotation of mass M_T, radius R_T, and mean coefficient of inertia α_T. The actual figure is replaced by the equivalent sphere of the same volume.

Conditions of hydrostatic equilibrium imply a balance between the pressure forces on the one hand and the gravitation and centrifugal forces on the other. Explicitly, at the distance R from the centre

$$\frac{\mathrm{d}p(R)}{\mathrm{d}R} = -\left[\frac{GM(R)}{R^2} - \tfrac{2}{3}\omega^2 R\right]\rho(R), \tag{23}$$

where $M(R)$ is the mass contained within the sphere of radius R. We might notice here that equation (23) is the form particular to our problem of the more general Bernoulli equation expressing momentum balance for static flows (Landau and Lifschitz, 1959)

$$V(R) + \int_0^R \frac{\mathrm{d}p}{\rho} = \text{constant}, \tag{24}$$

as the differentiation of equation (24) will show, remembering that the potential V refers to the gravitational force.

The density and mass are related through the expression for mass conservation

$$\frac{\mathrm{d}M(R)}{\mathrm{d}R} = 4\pi R^2 \rho(R). \tag{25}$$

The equations (22) and (23) will be supplemented by the equation of state of the material, relating pressure to density and other variables. Explicitly take the equation of state in the form

$$p = p(\rho, S, n_i) \tag{26}$$

where S is the specific entropy (entropy per unit volume) and n_i is the molar number of the ith chemical constituent. Let $z = R_T - R$ be the depth below the surface, then

$$\begin{aligned}\frac{\mathrm{d}p}{\mathrm{d}z} &= \left.\frac{\partial p}{\partial \rho}\right|_{s,ni}\left(\frac{\mathrm{d}p}{\mathrm{d}z}\right) + \left.\frac{\partial p}{\partial s}\right|_{\rho,ni}\left(\frac{\mathrm{d}s}{\mathrm{d}z}\right) + \sum_i \left.\frac{\partial p}{\partial n_i}\right|_{\rho,s}\left(\frac{\mathrm{d}n_i}{\mathrm{d}z}\right), \\ &= [g - \tfrac{2}{3}\omega^2(R_T - z)]\rho(z),\end{aligned} \tag{27}$$

where the second expression arises from equation (23) and $g = GM/R^2$ is the local acceleration due to gravity. The composite expression (27) is rearranged into an equation for the change of density with depth

$$\begin{aligned}\frac{\mathrm{d}p}{\mathrm{d}z} = g\rho\left(\left.\frac{\partial p}{\partial \rho}\right|_{s,ni}\right)^{-1} - \tfrac{2}{3}\omega^2(R_T - z)\rho\left(\left.\frac{\partial p}{\partial \rho}\right|_{s,ni}\right)^{-1} \\ - \left[\left.\frac{\partial p}{\partial s}\right|_{\rho,ni}\left(\frac{\mathrm{d}s}{\mathrm{d}z}\right) + \sum_i \left.\frac{\partial p}{\partial n_i}\right|_{\rho,s}\left(\frac{\mathrm{d}n_i}{\mathrm{d}z}\right)\right]\left(\left.\frac{\partial p}{\partial \rho}\right|_{s,ni}\right)^{-1}.\end{aligned} \tag{28}$$

This formula can be arranged on the basis of approximations of various kinds.

A. ADAMS-WILLIAMSON APPROXIMATION

Suppose the planetary interior is isentropic and chemically homogeneous; further, suppose the planet is not rotating. Equation (28) reduces to the simple form

$$\frac{\mathrm{d}p}{\mathrm{d}z} = g\rho\left(\left.\frac{\partial p}{\partial \rho}\right|_s\right)^{-1}. \tag{29}$$

The adiabatic bulk modulus K_s is defined according to

$$K_s = \rho\left.\frac{\partial p}{\partial \rho}\right|_s, \tag{30}$$

so that equation (29) becomes, using equation (16)

$$\frac{\mathrm{d}\rho}{\mathrm{d}z} = \frac{g\rho^2}{K_s} = \frac{g\rho^2}{K_T}(1 + \beta\gamma T)^{-1}. \tag{31}$$

The thermal expansion for many minerals is sufficiently small to be neglected for many purposes, i.e. we set $\beta = 0$ (see Table 5). No distinction is then drawn between the specific heats at constant volume and at constant pressure, and there will be no thermodynamic distinction between the adiabatic and isothermal bulk moduli according to equation (17). Consequently we can replace K_s by the isothermal modulus K_T in equation (31) to good approximation and appeal to seismic data through the formula (1). Such data lead to a knowledge of the function φ defined by

$$\varphi = \frac{K_T}{\rho} = V_p^2 - \tfrac{4}{3}V_s^2, \tag{31a}$$

so that equation (31) takes the form

$$\frac{\mathrm{d}\rho}{\mathrm{d}z} = \frac{g\rho}{\varphi} = \frac{GM\rho}{(R_T - z)^2\varphi}, \tag{32}$$

an equation from which the density profile with depth can be found by integration from a knowledge of $\varphi(z)$. Equation (32) was first used by Adams and Williamson (1923) for the preliminary determination of the density distribution within the Earth.

The method can be extended to include rotation. Equation (28) becomes now

$$\frac{\mathrm{d}\rho}{\mathrm{d}z} = \frac{GM\rho}{(R_T - z)^2\varphi}\left[1 - \frac{2}{3}\frac{\omega^2(R_T - z)^3}{GM}\right]. \tag{33}$$

The second term in the bracketed expression is always small (with the possible exception of Saturn: see Table 2) and can be regarded as a perturbation.

B. THERMAL EFFECTS

These arise from the first term in the bracket on the right hand side of equation (28), which is rearranged as follows, retaining the condition of chemical homogeneity. From thermodynamic arguments (Hoare, 1946),

$$\left.\frac{\partial p}{\partial s}\right|_\rho \left(\left.\frac{\partial p}{\partial \rho}\right|_s\right)^{-1} \frac{\mathrm{d}s}{\mathrm{d}z} = \left.\frac{\partial \rho}{\partial s}\right|_p \frac{\mathrm{d}s}{\mathrm{d}z}. \tag{34}$$

But

$$\left.\frac{\partial \rho}{\partial s}\right|_p = \left.\frac{\partial \rho}{\partial T}\right|_p \left.\frac{\partial T}{\partial s}\right|_p$$

and the adiabatic temperature gradient τ is defined by (see Appendix 4)

$$\tau = \left.\frac{\partial T}{\partial s}\right|_p \frac{\mathrm{d}s}{\mathrm{d}z} = \frac{\beta_p T}{\rho C_p}\frac{\mathrm{d}p}{\mathrm{d}z}, \tag{35}$$

where β_p and C_p are respectively the coefficient of thermal expansion and the specific heat both at constant pressure, and T is the local temperature. Remembering that β_p is defined by

$$\beta_p = -\frac{1}{\rho}\left.\frac{\partial \rho}{\partial T}\right|_p \tag{36}$$

we have finally

$$\left.\frac{\partial p}{\partial s}\right|_p \left(\left.\frac{\partial p}{\partial \rho}\right|_s\right)^{-1} \frac{\mathrm{d}s}{\mathrm{d}z} = \rho\beta_p\tau. \tag{37}$$

Equation (28) for a chemically homogeneous body becomes:

$$\frac{\mathrm{d}\rho}{\mathrm{d}z} = \frac{GM\rho}{(R_T - z)^2\varphi}\left[1 - \frac{2}{3}\frac{\omega^2(R_T - z)^3}{GM}\right] - \rho\beta_p\tau. \tag{38}$$

This modified form of the Adams-Williamson equation was derived first by Birch and later applied to terrestrial problems by Bullen (1936, 1963), though neglecting rotation.

C. CHEMICAL INHOMOGENEITIES

The treatment of inhomogeneities can be most complicated according to (28), but a measure of the importance of such effects can be obtained fairly easily. Equation (31) can be applied quite generally, and leads immediately to the relation

$$\frac{d\rho}{dz} = \frac{GM\rho}{(R_T - z)^2\varphi}\frac{\partial K}{\partial p} - \frac{\rho}{\varphi}\frac{\partial \varphi}{\partial z} = \frac{GM\rho}{(R_T - z)^2\varphi}\left[\frac{\partial K}{\partial p} - \frac{(R_T - z)^2}{GM}\frac{\partial \varphi}{\partial z}\right], \quad (39)$$

where we have used equation (23) but with $\omega = 0$ (no rotation). Compare equations (39) and (32). These equations become identical if

$$\frac{\partial K}{\partial p} - \frac{(R_T - z)^2}{GM}\frac{\partial \varphi}{\partial z} = 1, \quad (40)$$

when conditions are those of chemical homogeneity. For an inhomogeneous region the equality to unity will not apply and the bracket in the last term on the right in equation (39) has been used as a measure of deviation of a given region from conditions of homogeneity. This could be the result of a phase transition due to pressure. For use of the full equation (28) it is necessary to specify carefully the way in which the chemical composition of the planet varies with depth. It is seen from equations (35) and (36) that the effects of temperature variations and compositional vagaries act in opposite ways on the density. Consequently it can be anticipated that difficulties of interpretation can be expected in efforts to disentangle the effects of temperature variation and chemical inhomogeneities.

VII. Equations for Model Planets

The arguments of the last Section assumed a knowledge of the function φ [defined by equation (31a)] which can be deduced empirically as a function of the depth from seismic studies. The analysis can be developed alternatively on the basis of an assumed equation of state for the material (see Section IV) for application where seismic information is not available. The result will be a model planet of general validity, based on the properties of a class of constituent material.

A. DIMENSIONLESS FORM

In order to highlight the physical principles involved we will restrict our analysis to the simplest equation of state (6). Other equations could be used, such as the Birch-Murnaghan equation (11), but the additional analytical complications would be sufficiently great to hide much of the physics without introducing new principles (see Appendices 1 and 2).

It is convenient to introduce the following set of dimensionless variables, referring to the volume of the planet with the crust removed:

$$m = \frac{M}{M_T}, \qquad \lambda = \frac{M_c}{M_T}, \qquad r = \frac{R}{R_T}, \qquad \zeta = \frac{R_c}{R_T} \tag{41}$$

$$\hat{\rho}_i = \frac{3M_i}{4\pi R_i^{\,3}}, \qquad \bar{\rho}(r) = \frac{\rho(r)}{\hat{\rho}_T} = \frac{1}{3r^2}\frac{\mathrm{d}m(r)}{\mathrm{d}r} \tag{42}$$

$$\frac{\lambda}{\zeta^3} = \frac{\hat{\rho}_c}{\hat{\rho}_T} \equiv A, \qquad \alpha = \frac{I}{MR^2} \tag{43}$$

$$\theta_{Bi} = (4\pi)^{(B-1)}\left(\frac{G\rho_{0i}^{\;B}}{K_{0i}}\right)M_T(2-B)R_T(3B-4) \tag{44}$$

$$\chi = \frac{\omega^2 R_T^{\,3}}{GM_T}, \qquad \bar{p} = \frac{pR_T^{\,4}}{GM_T^{\,2}}. \tag{45}$$

The equations of hydrostatic equilibrium are (23), (35), (30), (25), and (6). It follows easily that the seismic function $\varphi = K_0 y^{(B-1)}$. In dimensionless variables the controlling equations are, therefore:

$$\frac{\mathrm{d}m}{\mathrm{d}r} = 3\bar{\rho}r^2, \tag{46}$$

$$\frac{\mathrm{d}\bar{\rho}}{\mathrm{d}r} = -\frac{3^{(1-B)}\theta_B}{r^2}\left[1 - \frac{2}{3}\frac{r^3}{m}\chi\right] - \bar{\rho}\tau\beta_p, \tag{47}$$

$$\frac{\mathrm{d}\alpha}{\mathrm{d}r} = 2\bar{\rho}r^4, \tag{48}$$

$$\frac{\mathrm{d}\bar{p}}{\mathrm{d}r} = -\frac{m}{r^2}\left[1 - \frac{2}{3}\frac{r^3}{m}\chi\right]\bar{\rho}(r). \tag{49}$$

These equations are to be solved subject to the boundary conditions:

$$m(0) = 0, \qquad m(\zeta) = \lambda, \tag{50a}$$

$$m(1) = 1, \tag{50b}$$

$$\alpha(1) = \alpha_T, \qquad p(1) = 0. \tag{50c}$$

These are a set of first order equations for determining the variables $\bar{\rho}$ and $\bar{p}$ for prescribed values of K_0 and ρ_0 which will generally change as the material composition changes with depth.

B. SECOND ORDER EQUATIONS

The first order equations (46)–(49) can be combined to form a second order equation for the density, the pressure or the mass. Such equations have limited interest in practice, but they are useful in showing some of the physics of planetary structures.

The equation for the mass is

$$\frac{\mathrm{d}}{\mathrm{d}r}\left(\frac{1}{r^2}\frac{\mathrm{d}m}{\mathrm{d}r}\right) + \theta_{Bi} r^{(2B-6)} m \left(\frac{\mathrm{d}m}{\mathrm{d}r}\right)^{(2-B)} \left[1 - \frac{2}{3}\frac{r^3}{m}\chi\right] - \frac{\hat{\rho}_T}{3r^4}\tau\beta_p \frac{\mathrm{d}m}{\mathrm{d}r} = 0. \tag{51}$$

This is a non-linear equation subject to the boundary conditions (50). With θ_{Bi} treated as either a constant or else only a slowly varying function of position, equation (51) is valid for regions of chemical homogeneity. Corresponding equations apply for $\bar{\rho}$ and $\bar{p}$ (see De Marcus, 1958, p. 421).

The simplest mathematical case is for an isothermal body (i.e. $\tau = 0$), when the third term on the left hand side of equation (51) vanishes. If θ_B is constant we can transform the variables r and m into the new independent variable t and dependent variable $s(t)$ according to

$$B = 1 + \frac{1}{n}, \qquad 3\bar{\rho} = s^n, \qquad t = \left(\frac{\theta}{n}\right)^{1/2} r. \tag{52}$$

Equation (51) is replaced by the second order equation

$$\frac{\mathrm{d}^2 s}{\mathrm{d}t^2} + \frac{2}{t}\frac{\mathrm{d}}{\mathrm{d}t} + s^n = 3\chi, \tag{53}$$

provided $B \neq 1$. For that special case (which is not of physical interest) we should need to introduce s according to $3\bar{\rho} = \ln s$, and the term s^n in equation

(53) would be replaced by exp (s). The boundary conditions (50) are replaced by the set

$$\begin{aligned} s(0) &= 0 & s^n(\zeta) &= 3\bar{\rho}_-(\zeta) \\ s^n(\zeta) &= 3\rho_+(\zeta) & s^n(1) &= 3\bar{\rho}_0 \end{aligned} \tag{54}$$

where the density $\bar{\rho}_-$ is to apply on the core side of the core mantle boundary and $\bar{\rho}_+$ is to apply on the mantle side of the boundary.

If $\chi = 0$ (non-rotating planet), equation (53) becomes the classical Lane-Emden equation of index n well known in the study of stellar interiors (Chandrasekhar, 1939). For stellar problems $n > 1$; for planetary problems $n < 1$. Generally, rotational effects are negligible except for Saturn; these effects are measured by the quantity χ, and numerical values for the planets are collected in Table 2.

The second order equation (53), or equivalently (51), has formed the starting point for several investigations. Thus equation (53) but with $\chi = 0$ has been used by Lyttleton (1965) to explore Mars, while Mars and the Moon have been treated using equation (51) but with $\beta = 0$ and $\chi = 0$ by Cole (1971) and Cole and Parkinson (1972).

1. *Effect of Parameter B*

The precise mathematical form of equation (53) depends upon the choice of B which characterizes the equation of state (6). It is easiest to consider first the case $B = 2$ which provides a linear form of equation (51). Restricting our arguments to non-rotating, isothermal conditions, the solution of equation (51) for $m(r)$ is a spherical Bessel's function of order $\frac{3}{2}$, suitably normalized (see Appendix 5). This is an oscillatory function with a period proportional to $\theta^{-1/2}$ but for planetary problems it is necessary to restrict the values of r to those less than the value for the first maximum: there is a maximum value for $\theta \sim \pi^2$.

Analogous arguments apply for values of B other than 2, although the mass distribution is not then represented by a Bessel's function. The form of $m(r)$ may be oscillatory for some values of B and monotonically increasing for others, but in all cases the restrictions implied by a maximum value of θ restrict the effects of the non-linearity on the calculated density distributions. Although the solutions of equation (51) appropriate to different values of B are very different over the full range of r, their forms are not very different in the restricted region between the origin and the first maximum which is the only range of interest now. These conclusions will be only slightly affected if account is taken of the phase change of the material due to pressure, or of chemical inhomogeneities.

C. OTHER EQUATIONS

More complicated equations than equation (51) will follow from the use of more complicated equations of state. Thus, the Birch-Murnaghan equation (8) leads to the expression for the compressibility function

$$\frac{\varphi}{\varphi_0} = \tfrac{3}{2}y^{2/3}\left[(\tfrac{7}{3}y^{2/3} - \tfrac{5}{3})(1 + \zeta) - \frac{\zeta}{3}y^{7/3}(9y^{2/3} - 7)\right]. \quad (55)$$

An equation for the density will follow from equation (32) which can be converted into an equation for the mass using equation (25). These expressions are too complicated to be set down here and we shall not need to refer to them explicitly.

It will be realized that discussions of this type, involving complicated equations of state, can hardly proceed without the use of computer facilities, and numerical studies are now a central part of the work in this field. The computational techniques are best suited to first order equations, such as the set (46)–(49) based on the equation of state (6), and second or higher order equations will generally not be of interest in this connection. Of particular interest are statistical methods possible involving Monte Carlo techniques (Press, 1971).

D. CONSTRUCTION OF MODELS

The equations developed so far have been used as the basis for the construction of models of various kinds. There are three broad approaches and we treat these separately.

1. *Using Seismic Data*

The internal conditions for the Earth have been extensively investigated using the various formulae of Section VI with φ deduced from seismic measurements (see Jeffreys, 1962; Bullen, 1963). The formulae, such as (32), are integrated from the surface inwards to the centre. It will be seen, from Section IX.A.1, that a very detailed picture of the terrestrial interior has been built up in this way. It will surely be only a matter of time before analogous data can be obtained first for the Moon and then for Mars.

A fine test of the correctness of a model can be developed by a comparison between calculated and measured data for the free oscillations (Bullen, 1965) and this is the trend of recent work (e.g. Dziewonski, 1971).

2. *Specification of Materials*

The most widely applied approach to general planetary problems is that using an equation of state for a prescribed material composition. The mean weighted density and bulk modulus at zero pressure are assigned as function of the depth, the equations of Section VI applied step by step from the surface of the mantle inwards. It must be realized that this is not the visible surface of the planet because the crust is not in hydrostatic equilibrium and so its mass, thickness and moment of inertia must be subtracted from the corresponding observed values for the planet as a whole. The pressure due to the crust (of the order of 10^9 dynes cm^{-2}) is the pressure at the surface for calculations [see equation (40c)] but this value is smaller than the central pressure by a factor of at least 10^{-2} (see Table 7) and could be neglected in preliminary

Table 7

Data referring to the centre of planetary bodies. p_{CL} is the minimum pressure calculated from equation (8.3), while p_c and ρ_c are respectively the central pressure and density calculated from the formulae of Section VII.

Planet	p_{CL} $\times 10^{11}$ dyne cm^{-2}	p_c $\times 10^{13}$ dyne cm^{-2}	ρ_c gm cm^{-3}
Mercury	3	0.6	8.3
Venus	12	3.3	12
Earth	17	4.0	13.1
Mars	1.6	0.6	8.5
Jupiter	100	11.0	31.0
Saturn	19	5.5	15.6
Uranus	18	4	~ 12
Neptune	36	5	~ 12

calculations. It is usually sufficient to suppose the crust to be a uniform shell of thickness h_1 and density ρ_1, since its contributions to the mass and moment of inertia of the planet as a whole are very small.

Models are constructed of the total planet of assigned radius, and the mass and coefficient of inertia calculated. Those models with values which correspond to a chosen planet are accepted as possible representations (as a representative sample of references see: Binder, 1969; Nakamura and Latham, 1969; Reynolds and Summers, 1969; Ringwood and Clark, 1971).

For the Murnaghan equation (6) the equations for the calculations are the set (46)–(49) or equivalently the second order equation (51). The equations for the Birch-Murnaghan equation (11) can be set down by the reader. For isothermal conditions we set $\beta = 0$, but if thermal effects are included it is necessary to assume a value of β. Although β is in general a function of pressure and temperature, the former effect is the more important and can be included through an expression of the form (Ringwood and Clark, 1971)

$$\beta = \beta_0\left[1 - \frac{4p}{K}\right]. \tag{56}$$

The temperature profile within the planet could be inferred from a knowledge of the thermal conductivity of the constituent material (both for conductive and radiative transfer processes) together with information about the internal distribution of heat sources (see Bullard, 1958, p. 110). This information is not known but an empirical profile can be constructed by fitting a power law formula of the type (Nakamura and Latham, 1969),

$$T(r) = T(0) - [T(0) - T(1)]r^n \tag{57}$$

for the temperature at a distance r from the centre. Here n is an index to be assigned. The heat flux through the surface, q, is given by

$$q = \lambda n[T(0) - T(1)] \tag{58}$$

for a given thermal conductivity λ and temperature difference between the surface and the centre, measured values of q allowing a value of n to be assigned. Usually it is sufficient to suppose $n \sim 4$, giving a strong gradient of temperature near the surface and an essentially uniform temperature in the central region. Alternatively, Reynolds and Summers (1969) suggest the usefulness of the empirical relation between temperature and pressure

$$T(r) = T(1) + p^{1/3}(r)T(0). \tag{59}$$

With $T(0)$ and $T(1)$ appropriately assigned this profile agrees acceptably well with other estimates.

The main computational criterion for the model is that the pressure should everywhere be continuous. If this is to be achieved, and the correct total mass and moment of inertia reproduced, it is often necessary for the density to be discontinuous at one or more points. This will be in addition to pressure-phase change effects, which will also need to be accounted for, for pressures in excess of a few times 10^{11} dyne cm^{-2}. By analogy with the Earth, it will be necessary to consider the possible existence of an iron core (remembering the arguments of Wiechert referred to in Section I) although there will usually not be a unique value for the core radius. A range of possible core configurations can be relevant although for reasons of stability the largest core size will be preferred. Of course such arguments presuppose a fully differentiated core, but the process of differentiation may well be far from complete with heavier materials still moving towards the central region (Urey, 1962). The possibilities are clearly very large. A mathematical technique designed for solving problems of this type is the Monte Carlo method (see Killingbeck and Cole, 1971) and it has been applied to the construction of terrestrial models by Press (1970, 1971). The results are highly interesting and can be arranged to give consistent information about chemical composition. [It is interesting to notice that the occurrence of earthquakes can also be usefully studied using Monte Carlo techniques (Knopoff, 1971).]

The core region is particularly difficult to treat because its contribution to the total mass and moment of inertia is small. For the Earth it is the region where seismic data are most unreliable; other methods are still awaited to make the analysis of this region precise, and the study of free oscillations may well be valuable here.

These arguments have been applied to the planets, including the Earth, separately and also collectively (see Section IX).

3. *Problem Inverted*

It is of interest to view planetary structures in general terms without presupposing particular constituent materials and this can be done to a limited extent in some cases, and particularly for an essentially homogeneous material. For definiteness suppose the equation of state (6) to apply so that the equations (46)–(50) are valid; for simplicity we restrict ourselves to an isothermal ($\beta = 0$), non-rotating ($\chi = 0$) planet. The calculated mass and moment of inertia of the planet can be adjusted by the appropriate choice of K_0 and ρ_0 for a chosen value of the state parameter B (this is clear from the equations of Appendix 5 as a special case). This method has been applied to the Moon (Cole, 1971) treated as a homogeneous body; it has also been extended to include an iron core (Cole and Parkinson, 1972) and in this form can be applied generally to planetary bodies. One use of the method is for the exploration of the core, the mantle constitution being presumed known. Another use is in connection with the Jovian planets, where the composition is open to considerable doubt and a preliminary analysis is valuable on the basis of a simplified model involving only a few chemical constituents.

VIII. Some Inequalities and Relations between Variables

Before undertaking detailed numerical calculations it is of use to have an indication of the orders of magnitude of variables, possibly expressed by upper and lower numerical bounds. Many of the arguments involve a mass weighted mean value $\bar{Q}$ of the variable Q defined by the integral

$$M_p\bar{Q} = \int_0^{M_p} Q \,\mathrm{d}M(r) = Q(R_p)M(R_p) - \int_0^{R_p} M \frac{\mathrm{d}Q}{\mathrm{d}M} \,\mathrm{d}M, \tag{60}$$

where, as usual, $M(r)$ is the mass contained within a sphere of radius r. As an example, for the pressure we set $Q \equiv p$; remembering that $p(R_p) = 0$, that $\frac{\mathrm{d}p}{\mathrm{d}M} = \frac{\mathrm{d}p}{\mathrm{d}r}\left(\frac{\mathrm{d}M}{\mathrm{d}r}\right)^{-1}$, and using equation (23) we find easily that

$$4\pi M_p\bar{p} = \int_0^M \frac{GM^2\,\mathrm{d}M}{r^4}\left[1 - \frac{2}{3}\chi\frac{M_T}{M(r)}\left(\frac{r}{R_T}\right)^2\right]. \tag{61}$$

For a body in hydrostatic equilibrium the density will not decrease inwards so that $(r^3/M(r)) \leqslant (R_p{}^3/M_p)$. Consequently, equation (61) is translated into the inequality

$$\bar{p} \geqslant \frac{1}{5}\left(\frac{4\pi}{3}\right)^{1/3} GM^{2/3}\bar{\rho}^{2/3}\left[1 - \frac{2}{3}\chi\right], \tag{62}$$

which provides a lower bound for the pressure. Of course, different methods of defining averages provide different values for the bounds (see Appendix 6).

Bounds for the central pressure p_c are readily found. On the one side the central pressure will be greater than that of the same mass but of constant density equal to the central density ρ_c, and therefore of smaller radius than the planet of interest. On the other side, p_c will be less than the pressure at the centre of the planet of the same radius but of constant density ρ_c. Each of these limits can be computed and the upper and lower bounds established. Relations involving the pressure can be converted into statements about the density through an equation of state. Values for p_c and ρ_c for the planets are collected in Table 7.

Other variables can be treated in comparable ways. Thus, if the planet has a solid surface, Jeffreys (1924) has shown that the surface density ρ_s satisfies the inequality

$$\rho_s \leqslant \tfrac{5}{2}\bar{\rho}_T\alpha_T, \tag{63}$$

a result later refined by De Marcus (1958). Values of ρ_s for the planets are collected in column 2 of Table 2. Bounds for the temperature cannot be determined because the thermal equation of state is unknown.

The methods apply equally to quantum systems through the equation of state and can be used to widen the scope of our discussion. For instance, for a non-relativistic Fermi gas the pressure is a function of the atomic weight and atomic number, according to equation (20), together with the density. Pressures within the bounds considered above are possible only for a ratio of (W/Z) of the order unity which means effectively for a hydrogen composition. It can be concluded that equations of state describing such extreme conditions will be of interest only for planets of density compatible with a primary hydrogen composition, which means for Jupiter and Saturn. Support for a substantial hydrogen composition for Jupiter and Saturn comes from the general theoretical study of hydrogen planets (de Marcus, 1951).

IX. Application to Planets

The various arguments of the preceding sections have been applied to the planets in various ways and this work continues. The initial hopes of relating observed planetary features to a unique chemical structure has probably to be accepted now as unattainable but a qualitative understanding of planetary interiors has been built up which can be progressively improved. In making the present survey it is convenient to treat the terrestrial planets separately from the Jovian group.

A. TERRESTRIAL PLANETS

The Earth has been studied the most closely of the members of this group and it is natural to regard this as the standard against which other planets can be judged.

1. *The Earth*

Apart from a knowledge of M_p, R_p, and α_p, there are two incontrovertible pieces of seismic information, viz: one is the existence of the Mohorovičić discontinuity marking a real boundary between geological crustal and mantle regions, at a mean depth of 30 km under continental regions although it may be only a few kilometres below the ocean bottom; the other is the existence of a closely spherical mantle-core boundary, with a core radius $R_c = 0.545 R_p$. Insignificant error is introduced by supposing the core-mantle boundary to be smooth and sharp, but it will not be exactly so and irregularities here could well be important in connection with the secular magnetic field but such considerations are outside the scope of this paper (Rikitake, 1971; Jacobs, 1971; Hide, 1969).

The density profile with depth has been calculated by many authors using the formulae of Section VI with φ assigned on the basis of P and S wave propagation: Bullen (1963) has been especially prominent in the work. The starting point for the inward integration of the equations of motion is taken at a depth of 33 km where the density is taken to be 3.32 on the basis of lava and meteoric evidence and by comparison with lunar material which is supposed similar to the upper mantle material. The density and pressure variations with depth are shown in Fig. 1, and other data are listed in Table 8.

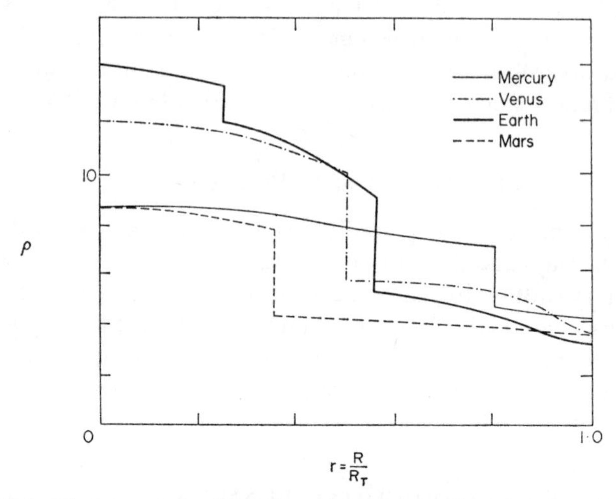

FIGURE 1. The internal density ρ (in units of gm cm^{-3}) as function of the depth for the terrestrial planets, calculated from the arguments of Section VII and assuming that each planet has a central core of material comparable to that of the Earth.

The Debye temperature can be found using equation (17) if the atomic weight of the material is known. The present best value for the Earth is probably $W = 25$ (Knopoff, 1965; Anderson *et al.*, 1968). On the basis of calculations such as these seven distinct regions (see Table 8) are recognized, including a distinction between the upper and lower mantle by different values of the pair of variables ($K_0 \times 10^{12}$, B). Using equation (6), the best values of these variables for the upper mantle are (1.2, 4.6) respectively while for the lower mantle the corresponding values are (2.3, 3.2). The transition region is thin and could be represented by a quadratic fit (Bullard, 1954, p. 93). Only the crustal region is not assumed to be in hydrostatic equilibrium.

Separately from assumed values of φ, Earth models can be constructed on the basis of an assumed equation of state such as (6), or (11), including those based on the methods of Section VII. Typical of this type of data are those collected in Table 8 on the basis of the linear relation between compressibility and pressure. For the mantle, B is taken to be 4 throughout; for the core it is 3.3. It may be preferable to suppose a quadratic relationship between K and p for the iron core material (Bullen, 1954, p. 244) but the distinction is not critical (see Appendix 1).

Table 8
Data for the interior of the Earth. The second column uses the arguments of Section VII,D,3 while the third column lists the data due to Jeffreys and Bullen (see Bullard, 1954, p. 93) derived from the formulae of Section VI and seismic data.

Variable	Methods of Section 7.4.3	Jeffreys-Bullen
$\rho(33)$ gm cm^{-3}	3.32	3.32
$K(33)$ dyne cm^{-2}	1.16×10^{12}	1.16×10^{12}
$V_p(33)$ km sec^{-1}	7.75	7.75
$V_s(33)$ km sec^{-1}	4.36	4.35
Θ °K	598	620
$\rho(500)$ gm cm^{-3}	4.09	3.89
$K(500)$ dyne cm^{-2}	1.77×10^{12}	2.15×10^{12}
$V_p(500)$ km sec^{-1}	8.65	9.66
$V_s(500)$ km sec^{-1}	4.85	5.32
$\rho(1000)$ gm cm^{-3}	4.53	4.68
$K(1000)$ dyne cm^{-2}	2.68×10^{12}	3.59×10^{12}
$V_p(1000)$ km sec^{-1}	10.09	11.43
$V_s(1000)$ km sec^{-1}	5.67	5.36
$\rho(2898)$ gm cm^{-3}	5.67	5.68
$K(2898)$ dyne cm^{-2}	6.56×10^{12}	6.5×10^{12}
$V(2898)$ km sec^{-1}	14.12	13.64
$V(2898)$ km sec^{-1}	7.93	7.30
$\rho(5121)$ gm cm^{-3}	11.70	14.20
$p(5121)$ dyne cm^{-2}	3.17×10^{12}	3.27×10^{12}
ρ_c gm cm^{-3}	13.07	17.20
P_c dyne cm^{-3}	3.55×10^{12}	3.64×10^{12}
β%	31.49	31.70
α_p	0.3335	0.3320

No more than a broad indication of likely minerals within the Earth can be expected to result from such calculations. There are three sources of information here; one is the observed solar abundances, relevant if the planets are supposed in some sense derived from the Sun; a second is from a study of meteoric material of various kinds assumed related to planetary formation; and the third involves surface evidence such as volcanoes. From such evidence it appears that 95% composition by mass is made up of Si, Mg, Al, Ca, Fe, Ni, and O. Of these, the first four appear principally as oxides, whereas Fe and Ni can exist as oxides or, deep inside near the centre, as free metals. It is more accurate to use generic terms than specific ones for terrestrial material and we follow the nomenclature of Reynolds and Summers (1969) in calling the oxides collectively rock, and free iron and nickel simply iron. The equations of state for rock and iron materials are constructed using the assumption (18) of addition of partial volumes of constituent materials, although this assumption is known to have only limited validity. This is particularly the case for pressures in excess of about 3×10^{11} dyne cm^{-2} (corresponding to a depth of 300 km in the Earth) where silicates can be expected to break down into component oxides which will then determine the densities. The location of phase rearrangement is open to uncertainties among which is the precise equation of state (see e.g. Bullard, 1954, p. 101). The distinction between the upper and lower mantles can be understood on this basis (Cook, 1970). Good qualitative information can, however, be achieved on a very simple basis such as that provided by equation (6) (see Table 8).

The core region is more difficult to explore. Three separate regions can be recognized, the outer two of which do not transmit transverse S-waves, although the central region may (Lehmann, 1934, 1953). The central *G*-region will be pure iron, the *E*-region outside could be an iron-silicates mixture but very heavily weighted to iron, while the *F*-region would be a transition region between the two. This complex region can be interpreted as being composed of a liquid outer (*E*, *F*) region and a solid central (*G*) region. These conditions can be viewed in terms of a complicated dependence of melting point on pressure (Jacobs, 1963) and the temperature distribution becomes of interest (Birch, 1965).

Several estimates of the internal temperature distribution have been attempted and are displayed in Fig. 2. The deduction of an analytic expression for the temperature distribution still has to be achieved but simple empirical expressions have been suggested. The expression (59) is one such proposal, the constants $T(1)$ and $T(0)$ being chosen to satisfy prescribed values, such as at the surface and lower down (e.g. at the core-mantle boundary). The pressure is found to be closely a linear function of the depth, and in this case equation (59) implies the relation $T^3 \alpha z$. This form agrees moderately with the distribution proposed by Tozer (1959). Indirect evidence can provide upper and lower bounds on the temperature at internal points; thus, the mantle temperature must be lower than the melting temperature of the rock under the appropriate local pressure, whereas the outer core temperature must be above the appropriate melting temperature locally although the inner core may need to be above it. A finer limit could perhaps be recognized when details of mantle and core convection processes become available. The main terrestrial magnetic field is now generally accepted as establishing the existence of convection currents in the outer core: continental drift and plate tectonics could imply the existence of convection currents within the mantle. The dynamics of such convection processes, and particularly the time scales, could through inference provide pointers to the temperature of the material when such motions are better understood (Elsasser, 1971; Gans, 1972).

The surface of the Earth is a constantly changing layer, covered by water over some three quarters of its area. Many surface rocks are younger, in

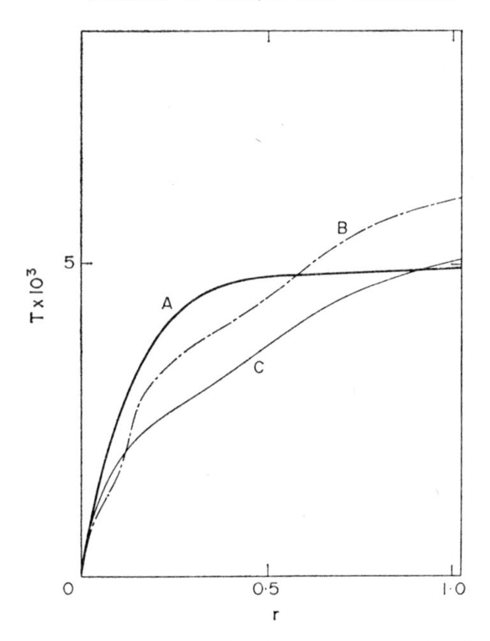

FIGURE 2. Inferred temperature distributions within the Earth. Curve A involves a model thermal history; curve B is based on data due to Tozer (1959); and curve C is the adiabatic problem based on the formula (A4.3).

their present condition, than 300 million years although the age of the Earth itself is in the general region 4.5–5 $\times$ 10^9 years. The Earth appears to be a highly differentiated body and this process appears still to be going on, as is witnessed by earthquakes, volcanoes, mountain building, isostatic adjustment and the production of sea-floor material from the mid-ocean ridges. There is here a picture of activity and the presence of water in terrestrial rocks is central to the maintenance of these processes (Wyllie, 1971). There is no evidence that the power of the thermal driving engine is abating. Future work must be directed more towards the study of the associated dynamical processes.

2. *The Moon*

Our knowledge of the outer regions of the Moon has grown enormously as a result of the Apollo lunar landings, but the deep interior must still be explored by inference (Hinners, 1971). Direct inferences come about from seismic studies; two of the three seismic stations that have been set up on the surface are still operating (Press, 1971). The lunar seismographs are designed to recognize three distinct types of events—moonquakes, meteoric impacts, and impacts produced artificially. The lunar seismic pattern is markedly different from that obtained on Earth. Perhaps the essential characteristic is an extremely long reverberation time which can be as large as 4 hr for man-made events. The longitudinal and transverse waves have both been recognized but surface waves (corresponding to the terrestrial Rayleigh and Love modes) have not been found, presumably due to the presence of scattering in lunar materials. But the long reverberation time implies a high resonance elastic quality factor (Q), possibly several thousand, in contrast to the Earth where the Q value is low, perhaps 110. The high Q value would be accounted for by the absence of volatiles in lunar rocks. Contiguous faces of rock grains would tend to weld together and dissipation by slipping would not occur. The outer few tens of kilometres of the Moon are probably controlled by meteoric impact events and this region could well act as a form of waveguide for the propagation of elastic energy.

The seismic data at present available suggest an increase of elastic velocity with depth, with a marked seismic discontinuity at a depth of 65 km. The top 30 km is likely to be especially fractured, the seismic energy propagation involving a diffusion mechanism rather than the normal wave form. Below 65 km the material seems more homogeneous, and this boundary can be identified as that separating mantle from crust. The Moon is largely aseismic, the energy released by moonquakes being only some 10^{-9} of that released by earthquakes. This suggests an internal state that is very much more relaxed than that for the Earth. The sources of the principal seismic activity lie in a shell between the depths of 800 and 1100 km (i.e. 0.54 and 0.36 of the total radius). The activity shows a monthly dependence linked to the lunar orbit. Both longitudinal and transverse seismic waves can propagate in the material above the 0.36 R level, but there is evidence that only the longitudinal P-waves are found below this level. This conclusion will require confirmation but would suggest that the material of the inner third of the Moon is not able to withstand shear, i.e. is liquid. The immediate implication is that the temperature there is high enough to melt the material. The low mean density of the Moon would seem to preclude the possibility of an iron core (but see later); a silicate mixture could, of course, have an unusually low melting point. A hot central region would lead to a relatively high temperature in the mantle which would be compatible with a high heat flow through the lunar surface. This is, indeed, one of the features of the Apollo sites, the measured heat flow being about one half that of the Earth; this is a surprisingly high heat flow for as small a body as the Moon, in terms of terrestrial type material composition. The study of the lunar interior temperature distribution is likely to provide surprises in the future. Lunar rocks studied so far are composed of the usual terrestrial elements but show radical structural differences due to the total absence of water. They are depleted in volatile materials and enhanced in refractory elements in comparison with the Earth and have every appearance of having originated from a partial melting or differentiation of parent magma. The presence of water reduces the solidification temperature considerably, and the absence of water from the Moon would be consistent with a low internal temperature for solid material.

In contrast to the Earth the lunar surface shows few signs of activity. Indeed, it seems that most of the surface is covered with rocks older than 3×10^9 years, and rocks with an age of the order of 4×10^9 years are common. There is no sign of appreciable volcanic activity now and the surface has every appearance of having remained unchanged (apart from external effects such as meteor impact, cosmic rays and thermal breaking) for the last 3×10^9 years. It seems likely that the lunar age is some 5×10^9 years, and it is now almost certain that the Moon was never physically attached to the Earth. Over 90% of the lunar surface is composed of continental type material, and is off-set in a way analogous to that of the continents and oceans on Earth. This could be interpreted in terms of internal lunar convection, but surface conditions might imply that such convection (if it ever existed) ceased some 3×10^9 years ago.

The data from the lunar magnetometer of the Apollo 12 mission has been coupled with that of the Explorer 35 orbit data to give some indication of the effect of the body of the Moon on the solar wind. This effect has been used to infer the distribution of the electrical conductivity with depth (Sonett *et al.*, 1971). Such a distribution can be associated with a temperature distribution and the present inference is that the internal temperature nowhere exceeds 1000°C. This could imply a solid internal condition throughout, a result that does not conflict with surface data but which might need reexamination in view of the seismic data.

The high value of the coefficient of inertia (the best value is possibly about 0.400 ± 0.004) would imply no central iron core. Three coefficients of inertia are necessary if the lunar motion is to be described with the greatest accuracy: one, α_p, refers to the pole as the axis, a second, α_E, to the line through the equator directed to the Earth, while the third, α_B, is perpendicular to the other two and parallel to the tangent to the orbit. These coefficients are in the ratios (Cook, 1972)

$$\alpha_p : \alpha_B : \alpha_E = 0.4032 : 0.4030 : 0.4029.$$

The condition of hydrostatic equilibrium cannot apparently be applied too rigorously to the Moon, although the deviations from this condition may not be very great. Present uncertainties of measurement would be consistent with a maximum iron core radius of $\frac{1}{10}R_p$, but it is likely that no core in fact exists (Cole and Parkinson, 1972). On this basis the Moon, quite unlike the

Earth, appears as an essentially undifferentiated body. This conclusion seems generally acceptable except for the extraordinary presence of remanent magnetism in the rocks which would imply the existence of a magnetic field within the Moon of magnitude about 10^{-2} G between the period about $3\text{–}4 \times 10^9$ years ago when the rocks solidified. The elucidation of this very puzzling feature must await data from later lunar missions. Magnetic data might imply the existence of a central core in spite of the large value of the coefficient of inertia. These could be explained if thermal effects are included, a not unreasonable temperature profile allowing the maximum assignable core radius to be increased to as much as $\frac{1}{5}R_p$.

3. *Mars*

Our knowledge of Mars is much more scant than that even for the Moon, but some firm conclusions are now becoming available. The photographic evidence of the surface recently obtained from Mariner 9 shows features which are recognizably terrestrial, mixed with other features which are lunar and others again which are typically Martian. This is quite different from earlier inferences (built up from terrestrial based observations) that Mars closely resembles the Earth in many respects. It is true that Mars has polar caps like the Earth, but these are now known to be composed of solid carbon dioxide rather than ice. Certain regions of the Martian surface contain volcanoes, but these can be more massive than those found on Earth (e.g. Nix Olympica has a diameter greater than 500 km at its base and is in excess of 15 km in height), volcanic caldera thirty times larger than in the Hawaiian Islands, and massive canyons (a series stretching east–west along the equator 80–120 km wide and 5 to 7 km deep are larger than corresponding terrestrial features). One feature that has caused particular interest is a number of channels that give every appearance of having been eroded by flowing water: but there is no other evidence to link water flows with the Martian surface. These channels extend out from a chaotic region which seems typically Martian. This region gives the appearance of a collapse, possibly related to the canyons observed to the west, but the nature of the collapse is not known.

Although such gross surface features appear in some areas, there are others that are correspondingly featureless. The nature of these areas is not clear; presumably they are young since otherwise they would show the results of meteoric impact. Dust storms can occur, and the storm that was in progress when Mariner 9 arrived in November 1971 covered much of the surface and lasted nearly 3 months. The atmospheric density cannot be anything like as dense as that of the Earth and the appearance of such substantial storms must imply a large source of very fine dust, with the consistency of face powder or finely ground cocoa.

The atmosphere of Mars is much less dense than was believed before the measurements of Mariner 9 became available. Thus the atmospheric surface pressure was found to be only about 1% that of the Earth even though with its gravitational pull Mars could hold an atmosphere whose pressure at the surface was ten times as great. But cloud structures have been observed that bear a close resemblance to terrestrial weather fronts.

One surface feature that presents considerable interest is the laminated structure surrounding each pole. Thus in the South polar region, in excess of thirty plates, as much as 200 km across and up to $\frac{1}{2}$ km in thickness have been made out and the North polar region is similar. The origin of these features is quite unknown but (since they seem unique to the poles) we may assume that there is a close connection with solid carbon dioxide depositions. A lack of craters in these regions suggests further that they are of relatively recent appearance. Quite generally, about one half the surface can be assigned as being of oceanic type materials (using the categorization familiar in terrestrial studies), the other half being continental (essentially the southern half). This approximate balance between the two types of surface material for Mars is to be contrasted with the unbalance for the cases of the Earth and the Moon.

Overall, the Martian surface gives the appearance of considerable activity presumably linked with corresponding internal movements. Erosion of surface features appears weak, suggesting that the atmosphere has never been dense. It would seem that Mars is only now heating up sufficiently to show tectonic changes and that much of its geological development is for the future. It is difficult to avoid the conclusion that it was colder in the past than now, and this could have implications for arguments involving the origin of the Solar System. As far as is known, Mars does not possess a general magnetic field, at least in excess of a few tens of gammas.

Models for the planet have been constructed by several authors (e.g. Jeffreys, 1964; Lyttleton, 1965; Binder, 1969; Reynolds and Summers, 1971; Cole and Parkinson, 1972) and all agree that the observed mean density and moment of inertia imply the presence of an iron core of about one-third the total planetary radius. This will carry a mass variously estimated between 5 and 12% of the total mass, values to be compared with a little less than 32% for the Earth. The internal temperature is quite unknown, but it is likely that the temperature exceeds the local solidus at least in the central regions.

Material can be expected to be obtained from the Martian surface in the near future from unmanned probes and it is foolish to speculate too widely on the likely form of such material. Polarization measurements (Dollfus, 1954) suggest that large areas will be found rich in iron oxide, and there is every reason to expect a broadly similar mineral content to the Earth. The presence or not of water will be a crucial measurement; tectonic features show the possibility of some recent surface movements but no definite conclusions can be drawn until measurements have been made *in situ* (for many details see Leighton *et al.*, 1971).

4. *Venus*

Of all the planets, Venus appears the most similar to the Earth, except for its substantially denser atmosphere and higher surface temperature. Unfortunately the moment of inertia of the planet is unknown and its virtual lack of rotation means that this information cannot be obtained from simple measurements. It is plausible to construct models on the basis of Earth models, using terrestrial type materials. This proves easy to do if an iron core is assumed with a radius of about $\frac{1}{2}R_p$, containing some 28% of the total mass. Such calculations hint at a lower iron content for Venus than for the Earth and suggest the possibility of a high degree of oxidation for Venus. Presumably the denser atmosphere suggests a higher degree of outgassing than for the Earth. Models of Venus based on the assumption of hydrostatic equilibrium can be used to infer a moment of inertia; the value found is generally around $\alpha_p \sim 0.341$ although no observational value is available for comparison.

5. *Mercury*

Very little is known about this planet and even its mean density is likely to be in doubt although it is certainly appreciably higher than that for Mars. The internal conditions are confused due to the planets close proximity to the Sun. As for the other planets so for Mercury, models can be constructed both with and without an iron core. If a core is present its radius must be some 0.8 of the total radius and it will contain some 68% of the total mass. The generally higher temperatures could imply a more reduced iron content than for the Earth. Such models provide the value $\alpha_p \sim 0.337$; the actual value is unknown.

B. JOVIAN PLANETS

These planets are characterized by large mass but low density, and small value for the coefficient of inertia. Presumably this must imply a large abundance of lighter elements coupled with a marked concentration of material towards the centre.

1. *Jupiter*

With a mean density almost identical to that of the Sun the fiery appearance of the planet in the telescope could easily be misinterpreted. But the thermal radiation is low and the planet is a cold condensed body. The Jeffreys–De Marcus inequality (see Section VIII) suggests the upper limit for the surface density 0.57 and we are clearly dealing with a different type of body from the terrestrial-type planet. The low surface density for Jupiter is consistent only with a composition of hydrogen and helium, and with the former as the major constituent.

Jupiter has a thick atmosphere which shows cloud bands in differential rotation, and it is natural to suppose hydrogen to be a major constituent, small proportions of other elements (such as nitrogen) existing in combination with hydrogen (e.g. methane). Remembering the low equivalent surface density (see Table 2) coupled with the low value of the moment of inertia and the high value of the surface gravity (some three times that of the Earth) we are led to suppose a thin hydrogen dominated isothermal atmosphere before the point of condensation is reached (see Wildt, 1961, p. 204). The atmosphere will then become first dense fluid hydrogen and then molecular liquid hydrogen without necessarily showing a clear surface of discontinuity (so that the terrestrial oceans do not provide an analogy), and at a greater depth will form solid hydrogen which in its turn will become metallic as we go still deeper. Below this, calculations based on the equations of Sections VII (and particularly Section VII,A,4) suggest the possibility of a central core of terrestrial type materials of radius $R_c = 0.15R_p$, possibly even including a central iron core. Models along these lines have been constructed by several authors and we show as a representative example in Fig. 3 data due to De Marcus. De Marcus has calculated the coefficients J_2 and J_4 in the expansion (A3.3) for the gravitational potential, which can be compared with values deduced from the observed orbits of the satellites. The values are

$$J_2\ (\text{calc.}) = 0.02130 \qquad J_2\ (\text{obs.}) = 0.02206$$
$$J_4\ (\text{calc.}) = 0.00250 \qquad J_4\ (\text{obs.}) = 0.00253$$

The comparison is sufficiently close to suggest that a model such as we have outlined is qualitatively correct, even though the quantitative details may be in error. The internal temperature is likely to be several thousand degrees at the centre.

Two features of Jupiter must be mentioned. One is the Red Spot whose nature is still as mysterious today as ever. The other is the existence of a permanent magnetic field which is either of great strength, if its origin is deep in the planet, or more likely has its origin in the shallower region of liquid hydrogen, presumably pressure ionized. There are indications from radio astronomy that the magnetic direction for Jupiter is reversed to the Earth, and that the apparent dipole field will make an angle of some 9° with the rotation axis.

2. *Saturn*

The picture to be constructed for Saturn is very much the same as for Jupiter but with two essential linked modifications. One is the extraordinarily

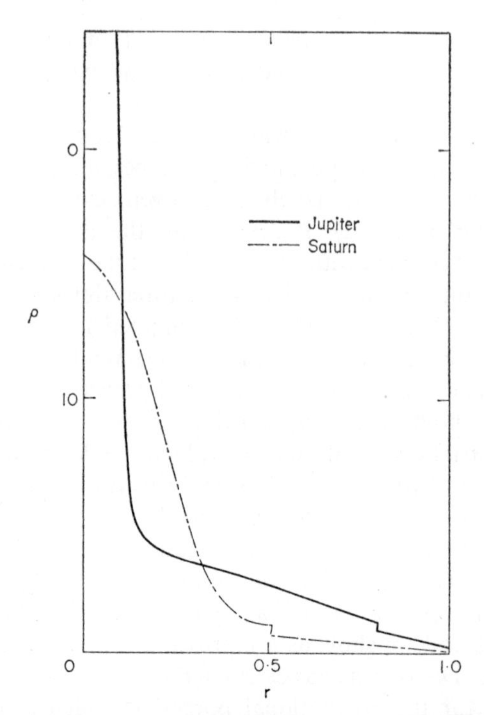

FIGURE 3. The internal density ρ (in units of gm cm^{-3}) as function of the depth for Jupiter and Saturn, calculated from the formulae of Section VII.

low values for the mean density, viz. 0.69, and the other is the low value of 0.22 for the coefficient of inertia. There is considerable difficulty here in producing a model with the appropriately low values of both of the quantities simultaneously unless account is taken of the rotation of the planet. When this is done the problem becomes tractable and the required percentage hydrogen content can be rather lower than for Jupiter. A representative model for Saturn is sketched in Fig. 3 from data given by De Marcus (1958). The coefficients in the expansion (A3.3) for the gravitational field for this model are

$$J_2(\text{calc.}) = 0.02594 \qquad J_2(\text{obs.}) = 0.02501$$

$$J_4(\text{calc.}) = 0.00588 \qquad J_4(\text{obs.}) = 0.00386$$

While the calculated value for J_2 is acceptable at this stage, the corresponding value for J_4 is reluctantly acceptable and further work is necessary. The characteristic and beautiful triple ring system of minute particles must be mentioned but discussion of its dynamical features and its possible origins are outside our present scope.

3. *Uranus and Neptune*

These planets present no particular difficulties from a computational point of view because their densities are relatively high (some ten times that of hydrogen configurations of the same mass), even though their coefficients of inertia are low (see Table 1). Models with a mantle and core can be constructed, and the inverse arguments outlined in Section VIII.D.3 are valuable here. The equations of state for hydrogen and helium are not of especial relevance now, but it is likely (Brown, 1950) that the properties of solid ammonia and methane will be of interest in calculations (Bernal and Massey, 1954). Again a central region rather larger than the Earth, but of broadly comparable material content, surrounded by an extensive light mantle can well provide the observed moments of inertia and masses of these two planets.

Conclusions

It is seen that the planets can be treated as a single unit within the established laws of the physics of cold bodies even though at first sight the terrestrial planets seem so radically different from the Jovian group. The planetary stability arises essentially from a balance between the gravitational and compressional forces, other effects such as those involving rotation or thermal energies playing only a secondary role. This gives the density and the bulk modulus locally a special significance in the theory.

The basic observational data are those made from the Earth using astronomical methods, space probes having improved the accuracy of some data but so far having provided nothing essentially new. One important exception is the lunar seismic stations and the data from this source can be expected to become increasingly detailed in the future. The number of parameters that are open to measurement is smaller in number than those necessary to specify the planetary interior even on a minimum basis and assuming hydrostatic equilibrium. When coupled with the inevitable errors of measurement, the result is an ambiguity of data that almost invites the exercise of personal predilections in any comprehensive description; this can be avoided if a suitable criterion of description is agreed upon. In particular it must be accepted as not possible to isolate a unique mineral composition for each planet but instead to agree a range of (ρ_0, K_0, B) values, compatible with the observational data and an equation of state, from which a range of planetary models, involving specific chemical compositions, can be constructed.

Viewing the planets on this basis the essential difference between the Jovian and terrestrial planetary groupings is the relative percentage of hydrogen content, the Jovian planets being closer to the Solar abundance than the terrestrial group. Within that framework, the Jovian planets appear to be composed of an extensive light mantle ($\rho_0 \sim 1$) and a denser central core ($\rho_0 \sim 6$) which could be composed of essentially terrestrial materials but with a radius something like double that of the Earth. Presumably this simple picture will change as more observational data become available, but already novel properties of matter arise, such as the existence of a metallic form of hydrogen. There is ambiguity for the terrestrial planets. One possibility is that the Earth alone has a heavy core, and if this be the case the mineral composition of each planet will be different and particularly the bulk modulus will be smaller by an order of magnitude than those values to be assigned to the terrestrial mantle. On the other hand, if the planets each have a core like the Earth's, a common mantle composition can be envisaged (with the possible exception of Mercury) although the relative proportions of materials will differ from planet to planet. If this were so the percentage of the total mass in the core would be about 28% for Venus, a little less than 32% for the Earth, about 12% for Mars and at most 1% for the Moon. Of course this is only a crude representation and each region would be expected to show considerable complication including a multiregion structure.

The arguments developed here involve the assumption of hydrostatic equilibrium and so exclude dynamical effects. Further developments in the subject can be expected to include dynamics. The existence of magnetic fields in the Earth and Jupiter imply the presence of convection currents at some point inside, and the changing surface appearance of the Earth (and on a more restricted scale of Mars) certainly imply the presence of slow mantle convection processes, possibly involving the entire mantle. Support for such a mechanism for the Earth can be claimed partly by the undoubted correlation between the fine details of the gravitational and magnetic fields and partly by the dynamics of plate tectonics that is only just beginning to be recognized. The future holds considerable excitement for the study of planetary interiors and the rewards can be great, not least of which could be a deeper understanding of the Earth itself.

Appendix 1

BIRCH-MURNAGHAN EQUATION

It is assumed that the Helmholtz free energy F is expanded in powers of the strain, $-f$, as

$$F = a_1 f^2 + a_2 f^3 + a_3 f^4 + \cdots. \quad \text{(A1.1)}$$

In this expansion the term linear in f is excluded by choosing the datum condition as that without strain, and the indeterminate constant always present in thermodynamic functions is set equal to zero. The strain f is defined by

$$\frac{\rho}{\rho_0} = \frac{V_0}{V} = (1 + 2f)^{3/2}. \quad \text{(A1.2)}$$

Thermodynamic arguments provide the definition of the pressure

$$p = -\frac{\partial F}{\partial V} = \frac{\partial F}{\partial f}\left(\frac{y^2}{V_0}\frac{\partial f}{\partial y}\right). \quad \text{(A1.3)}$$

Using the expressions (A1.1) and (A1.2) we find easily

$$p = \frac{2a_1}{3V_0} f(1 + 2f)^{5/2}\left(1 + \frac{3a_2}{2a_1} f + \cdots\right). \quad \text{(A1.4)}$$

Using the expression (3) for the compressibility, equation (8) results when it is realized that the zero-pressure compressibility is

$$K_0 = \frac{2a_1}{3V_0} \quad \text{(A1.5)}$$

and, when the function ξ is introduced according to

$$\xi = -\frac{3a_2}{4a_1}. \quad \text{(A1.6)}$$

Equation (A1.4) then transforms to equation (8) and equations (7) and (12) follow immediately.

The Birch-Murnaghan equation (11) is exactly true if the coefficient B_2 in the expansion (5) is related to B_1 and K_0 according to

$$\frac{\partial^2 K_0}{\partial p^2} = \frac{1}{K_0}\left[7\frac{\partial K_0}{\partial p} - \left(\frac{\partial K_0}{\partial p}\right)^2 - \frac{143}{9}\right]. \quad \text{(A1.7)}$$

The Murnaghan equation (6), on the other hand, requires

$$\frac{\partial^2 K_0}{\partial p^2} = 0 \quad \text{(A1.8)}$$

for its exact validity.

For a typical oxide, with $\dfrac{\partial K_0}{\partial p} \sim 4$, $K_0 \sim 10^{12}$ dyne cm^2, the right hand side of equation (A1.7) is of the order of magnitude 10^{-11}, which is sufficiently small to neglect. Consequently in practice, equations (6) and (11) lead to virtually the same equation of state data over a wide range of pressures. Where a distinction does arise, equation (11) is to be preferred.

Appendix 2

ALTERNATIVE EQUATIONS OF STATE

Equations of state other than that of Birch-Murnaghan (B-M) have been proposed and it is useful to collect these together here.

1. Reduced B-H forms.

For materials where $\xi = 0$, equation (11) is

$$p = \tfrac{3}{2}K_0(y^{7/3} - y^{5/3}). \tag{A2.1}$$

Alternatively, many terrestrial materials can be described by

$$1 - \xi(y^{2/3} - 1) = 2$$

so that then equation (11) is

$$p = 3K_0(y^{7/3} - y^{5/3}). \tag{A2.2}$$

These are essentially isothermal equations. The coefficient B_1 defined by equation (12) is different for the two cases: for equation (A2.1) we have $B_1 = 4$, while for equation (A2.2) we find instead

$$B_1 = 4 - (y^{2/3} - 1)^{-1}. \tag{A2.3}$$

This form cannot apply at the surface ($y = 1$); B_1 will change with depth.

2. There are two other isothermal equations involving two parameters which we list here.

(i) Born-Lande equation

$$p = \frac{3K_0}{(n-1)}\left[y^{(n+3)/3} - y^{4/3}\right] \tag{A2.4}$$

$$\frac{\partial K_0}{\partial p} = \frac{n+7}{3} \tag{A2.5}$$

which are compatible with the volume dependence of the Grüneisen parameter according to

$$\frac{\gamma}{\gamma_0} = y^{-n}. \tag{A2.6}$$

(ii) Born-Mayer equation

$$p = \frac{3K_0 y}{(1-2y)}\left[y^{-2}\exp\left(\frac{1-y}{y}\right) - y^{-4}\right] \tag{A2.7}$$

$$\frac{\partial K_0}{\partial p} = \frac{14 - (1 - 1/y)(2 + 1/y)}{6 - 3/y}. \tag{A2.8}$$

These equations, which differ in their derivation in the precise way in which lattice vibrations are accounted for, yield numerical data closely similar to each other and to the Murnaghan and Birch-Murnaghan equations of state (Thomsen and Anderson, 1969).

Appendix 3

SPHERICAL HARMONIC EXPANSION OF GRAVITY FIELD

Consider a planetary body in hydrostatic equilibrium in steady rotation such that the equilibrium ellipsoidal shape is closely spherical. Define an equivalent sphere of radius R_E chosen such that the sphere and ellipsoid have the same volume. Let Δ be a measure of the local deviation of the true figure from the ellipsoidal form (for the Earth the ellipsoidal form is the geoid), let ε be the ellipticity and η the radial distance from the centre of the planet to a point on the ellipsoid with colatitude θ. Then (Wildt, 1954; De Marcus, 1958: Jeffreys, 1962; Bullen, 1963):

$$\frac{\eta}{R_E} = \sum_{j\geqslant 0} a_j P_{2j}(\cos\theta) \tag{A3.1}$$

and

$$a_0 = 1 - \frac{4\varepsilon^2}{45} \tag{A3.2a}$$

$$a_1 = -\frac{2}{3}\left(\varepsilon + \frac{23}{42}\varepsilon^2 + \frac{4\Delta}{7}\right) \tag{A3.2b}$$

$$a_2 = \tfrac{8}{35}(\tfrac{3}{2}\varepsilon^2 + 4\Delta). \tag{A3.2c}$$

The potential V of the gravitational field external to the planet is expressed in an analogous way: explicitly

$$V = -\frac{GM_T}{\eta}\sum_{j\geqslant 0} J_j(\theta)\left(\frac{R_E}{\eta}\right)^j P_j(\cos\theta) \tag{A3.3}$$

$$J_j(\theta) = \frac{1}{M_T}\int\left(\frac{r}{\eta}\right)^j \rho(r)P_j(\cos\theta)\,\mathrm{d}\mathbf{r} \tag{A3.4}$$

and

$$J_0 = 1, \qquad J_1 = 0 \tag{A3.5a}$$

$$J_2 = \frac{I - A}{M_T R_E^{\,2}}, \tag{A3.5b}$$

and so on. Coefficients referring to higher order terms do not have a simple relationship to physically observable quantities.

Appendix 4

ADIABATIC THERMAL GRADIENT

An explicit expression for the adiabatic temperature gradient is obtained from thermodynamics, invoking if necessary the assumption of local equilibrium. For the movement of material in a gravitational field, the change of specific enthalpy, dH, due to a change in depth dz is given by

$$\mathrm{d}H = C_p\,\mathrm{d}T + (1 - \beta_p T)\frac{\mathrm{d}p}{\rho} + g\,\mathrm{d}z \tag{A4.1}$$

where dp and dT are the corresponding changes of pressure and temperature. For isenthalpic conditions, i.e. for d$H = 0$, the dependence of temperature on pressure (referred to as the Joule-Thomson coefficient, μ) is

$$\rho C_p \mu = \beta_p T + \frac{\mathrm{d}p^*}{\mathrm{d}p} - 1 \tag{A4.2}$$

where d$p^* = -\rho g$ dz is the pressure change for hydrostatic equilibrium.

If the motion of planetary magma and rock materials is very slow, hydrostatic equilibrium applied to a good first approximation. Then dp^* = dp and equation (A4.2) becomes

$$\rho C_p \mu = \beta_p T$$

or

$$\frac{\mathrm{d}T}{\mathrm{d}z} = \mu\frac{\mathrm{d}p}{\mathrm{d}z} = \frac{\beta_p T}{\rho C_p}\frac{\mathrm{d}p}{\mathrm{d}z} \tag{A4.3}$$

which is an expression for the adiabatic temperature gradient τ in terms of other variables.

The role of the Joule-Thomson coefficient in planetary internal dynamics has recently begun to be investigated.

Appendix 5

THE PARAMETER B IN EQUATION (4.4)

For an isothermal non-rotating planet, equation (51) becomes

$$\frac{\mathrm{d}}{\mathrm{d}r}\left(\frac{1}{r^2}\frac{\mathrm{d}m}{\mathrm{d}r}\right) + \theta_B r^{(2B-6)} m \left(\frac{\mathrm{d}m}{\mathrm{d}r}\right)^{(2-B)} = 0. \tag{A5.1}$$

This reduces to a linear form $B = 2$. Setting $\theta = \theta_2$, the solution of (A5.1) with $B = 2$ is

$$m(r) = \frac{\sin r\sqrt{\theta} - r\sqrt{\theta}\cos r\sqrt{\theta}}{\sin\sqrt{\theta} - \sqrt{\theta}\cos\sqrt{\theta}} \tag{A5.2}$$

which is often called the Laplace form. The density distribution and coefficient of inertia become respectively

$$\rho(r) = \hat{\rho}_T \frac{\theta \sin r\sqrt{\theta}}{3r[\sin\sqrt{\theta} - \sqrt{\theta}\cos\sqrt{\theta}]} \tag{A5.3}$$

$$\alpha_T = \frac{2}{3}\left[1 + \frac{6}{\theta\hat{\rho}_T}(\rho_0 - \hat{\rho}_T)\right]. \tag{A5.4}$$

The function θ is only weakly dependent on $\bar{\rho}_0$ and the quadratic expression

$$\alpha_T = 0.2618 + 0.1602\bar{\rho}_0 - 0.0220\bar{\rho}_0^{\,2} \tag{A5.5}$$

is a close numerical approximation to equation (A5.4).

It is seen from equation (A5.2) that $m(r)$ is a spherical Bessel's function of order $\frac{3}{2}$, i.e. $j_{\frac{3}{2}}(r\sqrt{\theta})$, suitably normalized. The value of α_T is smaller the smaller $\bar{\rho}_0$, but there is a lower limit to the density for a planet with a solid surface according to the Jeffreys expression (8.4) and so a lower limit to the moment of inertia of a homogeneous sphere composed of material satisfying the equation of state (4.4) with $B = 2$. Analogous arguments and conclusions follow from the use of other equations of state although they usually cannot be represented in closed mathematical form.

References

Adams, L. H. and Williamson, E. D., (1923). *J. Franklin Inst.* **195,** 475–529.

Ahrens, T. J., Lower, J. H., and Lagus, P. L. (1971). *J. Geophys. Res.* **76,** 518–528.

Anderson, O. L., Schreiber, E., Liebermann, R. C., and Soga, N. (1968). *Review of Geophysics*, **6,** 491–523.

Andrade, E. N. da C. (1947). Viscosity and Plasticity, Cambridge, Heffer.

Bernal, M. J. M. and Massey, H. S. W., (1954), *Geophys. Suppl. M.N.R.A.S.*, **114,** 172.

Binder, A. B., (1969), *J. Geophys. Res.*, **74,** 3110–3118.

Birch, F., (1952). *J. Geophys. Res.* **57,** 227–286.

Birch, F., (1961). *Geophys. J. R.A.S.*, **4,** 295–311.

Birch, F., (1965). *Geol. Soc. Am. Bull.*, **76,** 133–154.

Brown, H., (1950). *Astrophys. J.*, **111,** 641.

Bullard, Sir. E., (1958). *The Solar System* Vol. II (Ed. Kniper) Chapter 3 pp. 57–137 (Chicago Univ. Press).

Bullen, K. E., (1936). *Mon. Not. R. Ast. Soc., Geophys. Supp.*, **3,** 395–401.

Bullen, K. E., (1946). *Nature* (*Lond.*) **157,** 405.

Bullen, K. E., (1949). *Geophys. Suppl. M.N.R.A.S.*, **5,** 355.

Bullen, K. E., (1963). Introduction to Seismology. Cambridge.
Chandrasekhar, S., (1939). Study of Stellar Structure, Dover.
Chandrasekhar, S., (1961), Hydrodynamic and Hydromagnetic Stability (Oxford: Clarendon Press).
Chapman, S. and Bartels, J., (1951). Geophyics, Clarendon, Oxford.
Cole, G. H. A., (1971). *Planet. Space Science*, **19**, 929–947.
Cole, G. H. A. and Parkinson, D., (1972). *Planet. Space Science*, **20**, 557–569.
Cook, A. H., (1970). *Nature*, *Lond.* **226**, 18–20.
Cook, A. H., (1971). Gravity and the Earth. Wykeham Science Series.
Cook, A. H., (1972). *Proc. R. Soc.*, **A238**, 301–336.
Cowling, T. G., (1956). Magnetohydrodynamics, Interscience.
Darwin, G. H., (1899). *Mon. Notices R. Astro. Soc.* (*London*) **60**, 82.
Davies, G. F. and Anderson, D. L., (1971). *J. Geophys. Res.*, **76**, 2617–2627.
De Marcus, W. C., (1958). Handbuch der Physik, **58**, 419–448.
Dollfus, A., 1954, Solar System Vol. III (Ed. Kuiper and Middlehurst), p. 343. Chicago University Press.
De Sitter, W., 1924. *Bull. Astronom. Inst. Netherl.* **2**, 97.
Dziewonski, A. M., 1971, *J. Geophys. Res.*, **76**, 2587–2601.
Elsasser, W. M. (1971a). *J. Geophys. Res.*, **76**, 1101–1112.
Elsasser, W. M. (1971b). *J. Geophys. Res.* 4744–4753.
Franck, F. C., (1968). *Nature Lond.* **220**, 350–352.
Gilvarry, J. J., (1969). The Application of modern physics to the Earth and Planetary Interiors (Ed. S. K. Runcorn) pp. 313–403 (London, Wiley).
Hide, R. (1969). *Nature*, *Lond.* **222**, 1055–1057.
Hinners, N. W., (1971). *Rev. Geophys. Space Phys.* **9**, 447–522.
Hoare, G., (1946). Thermodynamics, Arnold, London.
Jacobs, J. A., (1971). *Nature*, *Lond.* **231**, 170–171.
Jeffreys, Sir H., (1924). *M.N.R.A.S.*, **84**, 534.
Jeffreys, Sir H., (1962). The Earth (4th Edition) Cambridge.
Joos, G., (1964). Theoretical Physics, Blackie, London.
Killingbeck, J. P. and Cole, G. H. A., (1971). Mathematical Techniques and Physical Applications, Academic Press, London and New York.
Knopoff, L. and MacDonald, G. F. J., (1958). *Geophys. J. R. Ast. Soc.* **1**, 284–297.
Knopoff, L. (1971). *Rev. Geophys. Space Phys.*, **9**, 175–188.
Knopoff, L. and Uffen, R. J., (1954). *J. Geophys. Res.*, **59**, 471–485
Lehmann, I., (1934). *Geodact. Inst. Skr.*, **5**, 44p.
Lehmann, I., (1953). *Bull. Seis. Soc. Amer.*, **43**, 291–306.
Leighton, R. B. and Murray, B. C. (1971). *J. Geophys. Res.*, **76**, 293–472.
Lodge, A. S., (1964). Elastic Liquids, Academic Press, London and New York.
Lyttleton, R. A., (1965). *Mon. Notices R. Astro. Soc.*, **129**, 21.
McQueen, R. G., Marsh, S. P., and Fritz, J. N., 1967, *J. Geophys. Res.*, **72**, 4999–5036.
Murnaghan, F. D. (1944). *Proc. Nat. Acad. Sci. U.S.A.* **30**, 244–246.
Murnaghan, F. D. (1951). Finite Deformation of an Elastic Solid, Wiley, New York.
Nakamura, Y. and Latham, G. V., (1969). *J. Geophys. Res.*, **74**, 3771–3780.
Orowan, E., (1965). *Trans. R. Soc. Lond.*, **A258**, 284–313.
Press, F., (1971). *Q. Jl. R.A.S.* **12**, 232–243.
Radau, R. (1885). *C.R. Acad. Sic.*, *Paris* **100**, 972.
Reynolds, R. T and Summers, A. L., (1969). *J. Geophys. Res.*, **74**, 2494–2511.

Rikotake, T. (1971). Essays in Physics Vol. 3, pp. 117–169. Academic Press, London and New York.

Ringwood, A. E. and Clark, S. P. (1971). *Nature, Lond.* **234,** 89–92.

Ryabinin, Yu. N., Beresnev, B. I., and Martinov, E. D., (1971). *J. Geophys. Res.* **76,** 1370–1375.

Sato, R. and Espinosa, A. F., (1967a). *Bull. Seismol. Soc. America*, **57,** 829–856.

Seidelmann, P. K., Klepcynski, W. J., and Duncombe, R. L., (1971). *Astroph. J.* **76,** 483–492.

Simcox, L. N. and March, N. H., (1962). *Proc. Phys. Soc. (Lond.)*, **80**, 830.

Stewart, J. W., (1956). *J. Phys. Chem. Solids*, **1,** 146.

Tayler, R. J., (1968). Structure and Evolution of the Stars, Wykeham Science Series.

Thomsen, L., (1971). *J. Geophys. Res.*, **76,** 1342–1348.

Thomsen, L. and Anderson, O. L., (1969). *J. Geophys. Res.*, **74,** 981–991.

Tozer, D. C., (1959). Physics and Chemistry of the Earth, Vol. 3 (Ed. Ahrens, Press, Rankama and Runcorn) Pergamon, New York.

Urey, H. C., (1952). The Planets, Yale Univ. Press, New Haven.

Wiechert, E., (1897). *Nachr. Ges. Wiss. Göttingen*, 221–243.

Wiechert, E., (1907). *Nachr. Ges. Wiss. Göttingen*, 415–529.

Wildt, R., (1954). The Solar System Vol. III (Ed. Kuiper and Middlehurst) Chapter 5, pp. 159–212, Chicago Univ. Press.

Woolfson, M. M., (1969). *Rep. Prog. Phys.* **32,** 135–185 London: Institute of Physics.

Wyllie, P. W., (1971). *J. Geophys. Res.*, **76,** 1328–1338.

Critical Phenomena

D. SETTE

Istituto di Fisica, Facoltà di Ingegneria, Università di Roma, Roma, Italy, and Gruppo Nazionale di Struttura della Materia del Consiglio Nazionale delle Ricerche, Roma, Italy

I. Phase-Transitions and Critical Points

The equilibrium of a thermodynamical system of particles at given values of the independent state variables corresponds to a rather complex interplay of tendencies determined by forces among particles, interactions with the external medium, and the statistical nature of the equilibrium itself. This is usefully expressed in thermodynamics by simple conditions valid for a suitable potential function. If for instance the system is held at constant pressure the equilibrium condition corresponds to the minimum possible value of the Gibbs free energy among various configurations of the system. If we consider for simplicity, the case of a fluid the Gibbs function is

$$G(p, T) = U - TS + pV \tag{1}$$

with p pressure, T temperature, U internal energy, S entropy, and V volume and, referred to one mole, it equals the chemical potential $\mu(p, T)$. Figure 1

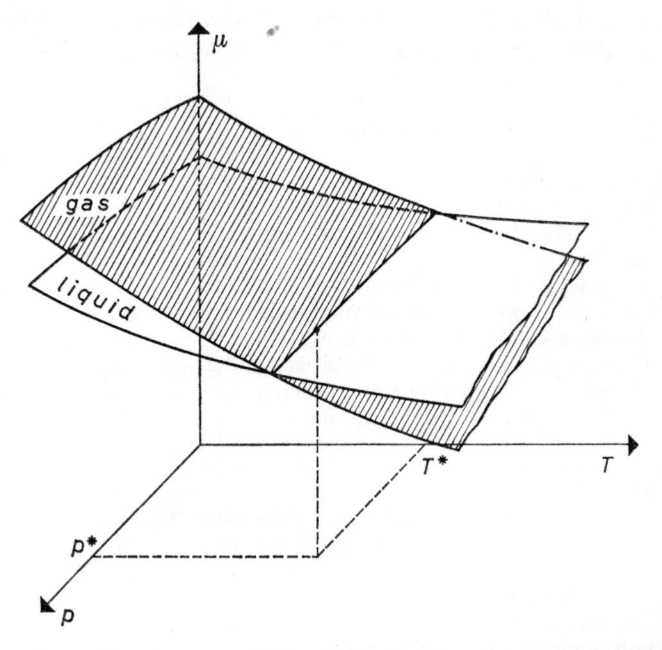

FIGURE 1. Chemical potential (or Gibbs potential) surfaces for two phases.

shows typical shapes of μ (or G) for the liquid and gaseous phases of a chemical substance in the μ (or G), p, T space: at a given pressure, p^*, the liquid phase is stable below a certain temperature T^*; the gaseous phase is stable at temperatures higher than T^*; the two phases are in equilibrium at T^*. At such a temperature, with a finite exchange of heat with the external medium, the system may pass abruptly from the characteristic structure of one phase to that of the other. The change of phase is accompanied by discontinuous variations of the first order derivatives of the thermodynamical potential, i.e. the volume $V = \left(\frac{\partial G}{\partial p}\right)_T$ and the entropy $S = -\left(\frac{\partial G}{\partial T}\right)_p$, as is evident from Fig. 1. Phase transistions of this kind, i.e. where one or more first order derivatives of the relevant thermodynamic potential change discontinuously, are called *first order transitions*. Continuing, for simplicity in introducing concepts and terminology, to consider the case of fluid substances, we find that the differences in properties of the system between phases at the transition temperature, decrease gradually as the equilibrium temperature increases (of course by imposing higher external pressures) until a temperature is reached when all differences disappear. In the p, T plane (Fig. 2) this is shown by the end of the line of equilibrium states between liquid and gaseous

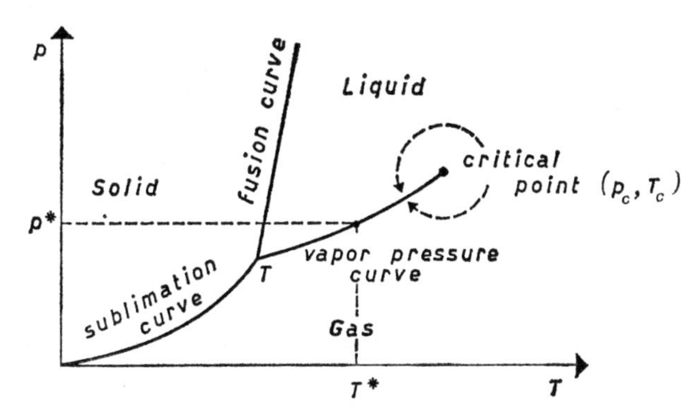

FIGURE 2. Phase equilibrium diagram for a simple substance.

phases. The equilibrium state reached is called *critical*, or, with reference to the equilibrium line, *critical point*; critical temperature (T_c) and critical pressure (p_c) are the corresponding values of the independent state variables.

The particular shape of the gas-liquid equilibrium line allows fluids, by suitable changes of temperature and pressure (e.g. along the dotted line in Fig. 2), to pass with continuity from states which have all the characteristic features of a liquid to states which are definitely gaseous (continuity of liquid and gaseous states); only at temperatures lower than the critical one is it possible to have the two phases (liquid and gas) simultaneously present in equilibrium.

Many other phenomena show the existence of a transition point, with many features analogous to those valid for the critical point of a one component fluid. In all cases there is a striking difference in the behaviour of the system when the temperature is changed through the critical point. While on one side two distinct phases may exist and the passage from the one to the other can occur by transformation of any quantity, on the other side only one homogeneous phase exists. At the critical point all changes are continuous and involve the whole mass of the system.

Examples of these phenomena are: (a) the phase separation in a binary liquid mixture or in binary metallic alloy—below (or above) a given temperature two distinct phases may exist while above (or below) the critical temperature a homogeneous solution exists at any composition; (b) the spontaneous magnetization of a ferromagnet, i.e. the existence of two differently oriented magnetic domains—any difference disappears at a critical (Curie) point; (c) the antiferromagnetic spin ordering and the existence of two opposing domains—it disappears at the Néel point; (d) superfluidity in He^4—below a critical point (λ point) distinct regimes of superfluid and normal flows exist, while all superfluidity disappears above; (e) superconductivity—below a critical point in a metallic superconductor all resistance disappears.

The transitions occurring in the way just outlined and for which a critical point (or better, in some cases, a line of critical points) is to be considered, are characterized by the fact that, in contrast with first order transitions, the first order derivatives of the relevant thermodynamic potential remain continuous at the critical point and only higher order derivatives (compressibility, specific heats, susceptibility) change discontinuously or show divergencies. These transitions are called *higher order transitions* or, as suggested by Fisher (1967), *continuous transitions* (with reference to the behaviour of first order derivatives of the thermodynamic potential). The processes occurring in the proximity of critical points, usually called critical phenomena, show many striking analogies and similarities, notwithstanding the different natures of systems and of quantities observed. This points to the existence in all cases of some common basic aspects in the physics determining the observed behaviour. It will be found that the critical phenomena are co-operative processes in the sense that they are produced by the mutual interactions of many particles and that an increase of microscopic fluctuations of suitable quantities occurring in the neighbourhood of the critical point is the common feature of paramount importance.

In the following paragraphs we will discuss critical phenomena by considering first the equilibrium properties (Section II) and then the dynamic behaviour (Section III). Of course, our treatment is far from exhaustive but, we believe, it to be adequate to furnish a general picture of the present status of knowledge and to illustrate the main directions of contemporary research. In particular after a brief review of some experimental results on equilibrium properties of many kinds of critical systems, we shall be concerned mainly with fluids and ferromagnets. Other information can be obtained from excellent general papers and books existing in literature and covering the whole field (Green and Sengers, 1966; Fisher, 1967; Heller, 1967; Kadanoff *et al.* 1967; Egelstaff-Ring, 1968; Sette, 1970; Stanley 1971) or specialized topics. Some of the latter will be indicated in the appropriate paragraphs. In these papers a much more extensive coverage of the scientific literature than is possible, or desirable here, is given. In the present paper, it is not our purpose to present an exhaustive account of the literature specifying in detail the relevance of each contribution. Rather, we wish to concentrate on essential issues.

II. Equilibrium Properties

A. SOME EXPERIMENTAL RESULTS IN CRITICAL PHENOMENA OF VARIOUS KINDS

In the present section the experimental evidence covering the main features of equilibrium critical phenomena will be given for systems of different types.

(1) *One component fluid system.* The first evidence of a critical point (as well as the name itself) is due to Andrews (1869), who studied the CO_2 system in a wide range of pressures and temperatures. The particular behaviour found for the isotherms of one component fluid system has furnished the basis of such important theoretical developments as the Van der Waals' state equation. Figure 3 gives similar isotherms in Xenon (Habgood and Schneider, 1954). In the low temperature region two separate phases exist and their densities (ρ_L for the liquid, ρ_G for the gas) change along the co-existence curve, the critical point is at the summit of the rounded coexistence curve where all differences between the two phases disappear. The first feature of interest is the behaviour of isothermal compressibility

$$K_T = -\frac{1}{V}\left(\frac{\partial V}{\partial p}\right)_T = \frac{1}{\rho}\left(\frac{\partial \rho}{\partial p}\right)_T \tag{2}$$

and this is readily obtained from the slope of the geometrical tangent to the isotherm at the point representing the thermodynamic state. This slope is non-zero in the one phase region except at the critical point. Figure 4 gives the general behaviour of K_T for $\rho = \rho_c$ in the one phase region ($T > T_c$) and for ρ_L and ρ_G in the co-existence region ($T < T_c$). The experimental values found for K_T in the proximity of T_c reach several million times the value

$$K_I = \frac{1}{\rho_c k_B T_c}$$

(k_B, Boltzmann constant) which is the compressibility that an ideal gas of density ρ_c would possess. *K_T diverges to infinity as $T \to T_c$.*

The high value of compressibility in the critical region leads to sizeable variations of densities for very small changes of pressure, like those produced by the weight of the fluid itself in a test tube of non-infinitesimal height. The

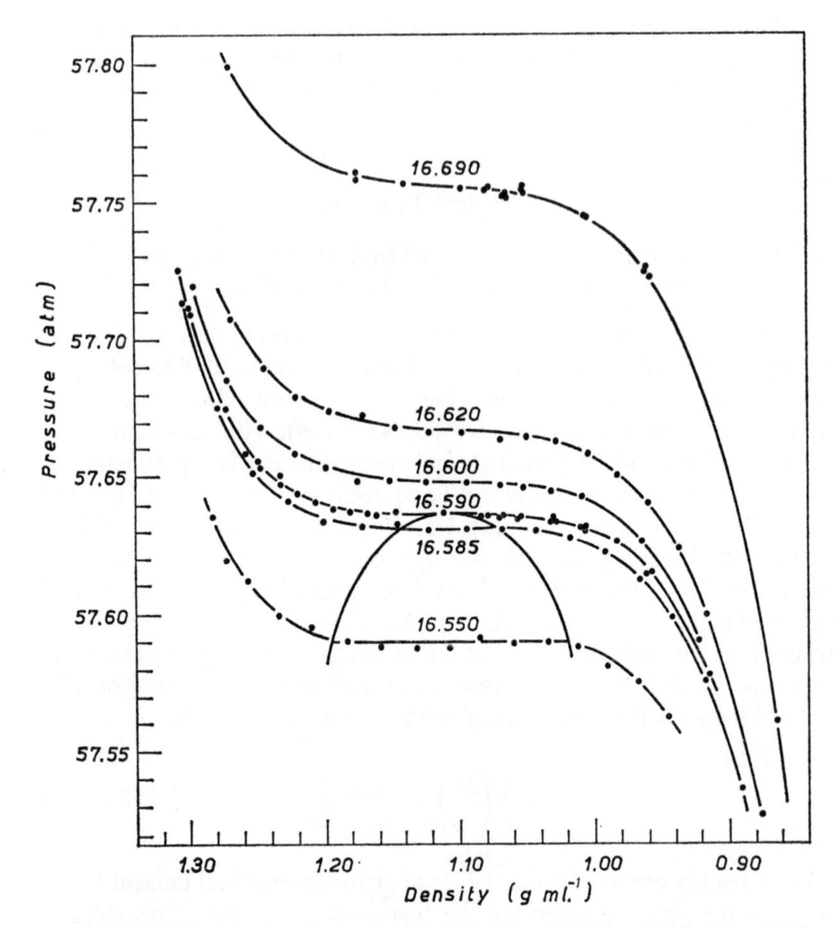

FIGURE 3. Pressure-density isotherms in Xenon (Hadgood-Schneider 1954).

existence of this *gravitational effect* has been responsible, when overlooked, for many faulty results on the shape of the co-existence curve; however, when rightly used (Lorentzen, 1953), it has allowed the best available results on the (p, ρ) isotherm near the critical point and the sharpness of the critical point itself.

The detailed shape of the critical isotherm is a second aspect of the behaviour of these systems which is worth considering. The results are usually fitted with an expression of the type

$$\left| \frac{p - p_c}{p_I} \right| = \text{const} \left| \frac{\rho - \rho_c}{\rho} \right|^{\delta}. \tag{3}$$

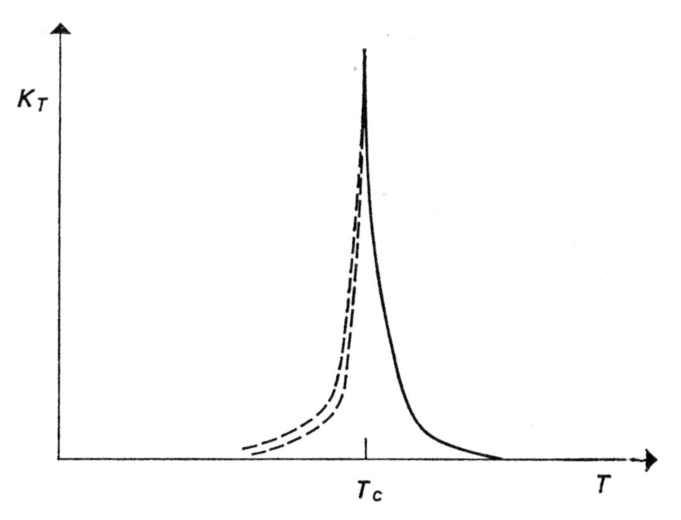

FIGURE 4. Isothermal compressibility in a fluid.

A third element of importance in the analysis of the behaviour of these systems is the shape of the co-existence curve. An interesting experimental result is that the shape seems to be the same for all one component fluid systems. Figure 5 gives the results up to 1945 as collected by Guggenheim (1945). The data are fitted with a mathematical expression of the type

$$\frac{\rho_L(T) - \rho_G(T)}{2\rho_c} = \text{const}\left(1 - \frac{T}{T_c}\right)^{\beta} \tag{4}$$

and the value $\beta \simeq \frac{1}{3}$ is indicated by experiments. Figure 6 gives the results obtained in CO_2 by Lorentzen (1965) using a method which profits from the gravitational effect. The differences $(\rho_L - \rho_G)$ are plotted as a function of $(\Delta T)^{1/3} = (T - T_c)^{1/3}$.

One can also consider $(\rho_L - \rho_G)$ and $(\rho_c - \rho_G)$. Like $(\rho_L - \rho_G)$, they are both proportional to $\left(1 - \frac{T}{T_c}\right)^{\beta}$. The quantity $(\rho - \rho_c)$ is the so-called "order parameter" which, as we shall explain later, has been introduced to measure the kind and amount of order that exists in the neighbourhood of the critical point.† It follows, therefore, that

$$\rho - \rho_c \sim \left(1 - \frac{T}{T_c}\right)^{\beta}. \tag{5}$$

† It is zero above T_c while it can assume two values below T_c—a positive one for the liquid phase and a negative one for the gaseous phase.

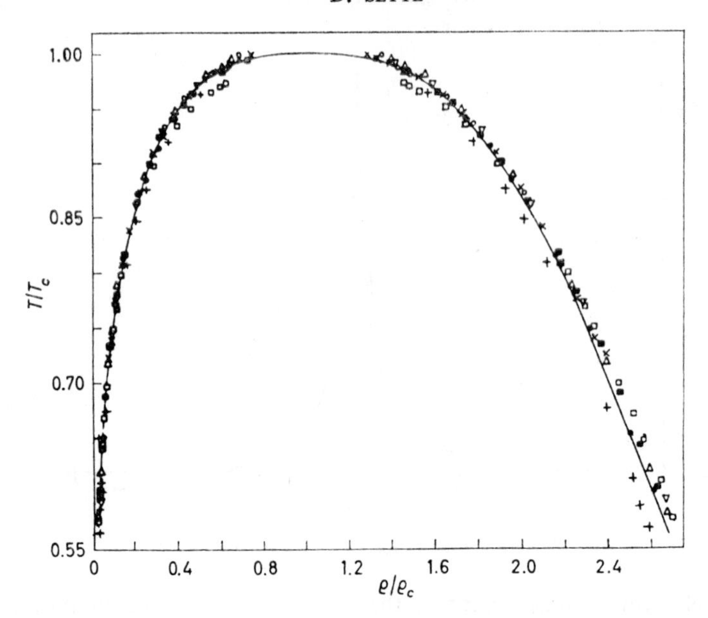

FIGURE 5. Experimental data on the co-existence curve of one-component systems (Guggenheim 1945).

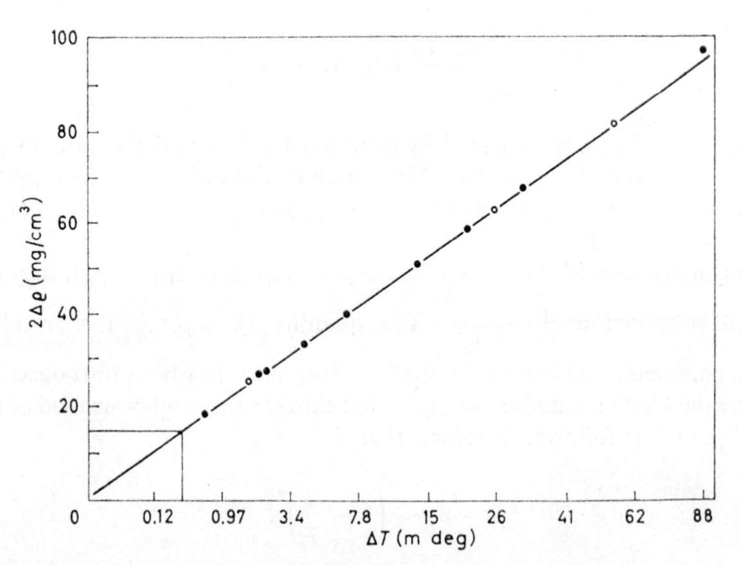

FIGURE 6. Coexistence curve for CO_2: difference density between liquid and gas phases vs $\frac{1}{3}$ power of $\Delta T = T - T_c$ (Lorentzen 1965).

A fourth critical property is the behaviour of the specific heat at constant volume, C_v. For a long time it has been assumed, on the basis of thermodynamic considerations, that $C_v(T)$ along the critical isochore had a discontinuity but remained finite as $T \to T_c$. The first measurements of C_v made with a high resolution method have however contradicted this result and instead have suggested a divergence of C_v along the isochore as the critical point is approached from either direction. Figure 7 shows the results

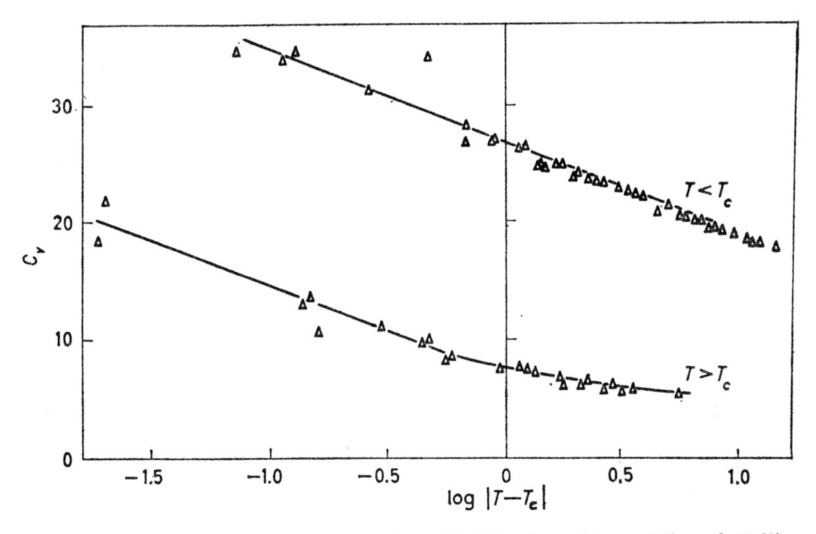

FIGURE 7. C_v for Argon along the critical isochore (Begastskii *et al.* 1962).

of calorimetric measurements in argon performed by Begastskii *et al.* (1962), where C_v is given as a function of $\log |T - T_c|$.

Still another striking experimental result furnished by a critical one component system is the characteristic *opalescence*. If a light beam is sent through the fluid when far from the critical region, only molecular scattering is observed; if however the fluid is brought into the critical region, an intense forward scattering appears and the whole volume of the fluid appears to glow. This process indicates the existence of large fluctuations of the refractive index, and of the local density. The occurrence of these large fluctuations is connected with the low values of the compressibility in the critical region. The work needed to compress a small region in the fluid is proportional to $1/K_T$ and when it becomes comparable with the thermal energy, $k_B T$, the region has a high probability of being compressed; this happens near T_c.

The accurate analysis of light scattering is a powerful tool for studying the critical phenomena. While the forward scattering may furnish information on the compressibility, the angular dependence as a function of temperature gives informations on the spatial dimensions of the fluctuations. These become larger and larger as the critical point is approached. It is possible also to have information on the time scale of fluctuations: they become slower and slower as $T \to T_c$ (critical slowing-down). This fact is responsible for the very long time which is required for the attainment of equilibrium in the critical region and special precautions are necessary for reliable results.

(2) *Binary liquid mixtures.* Some binary liquid systems are characterized by partial solubility of the two components (A and B) below (sometimes

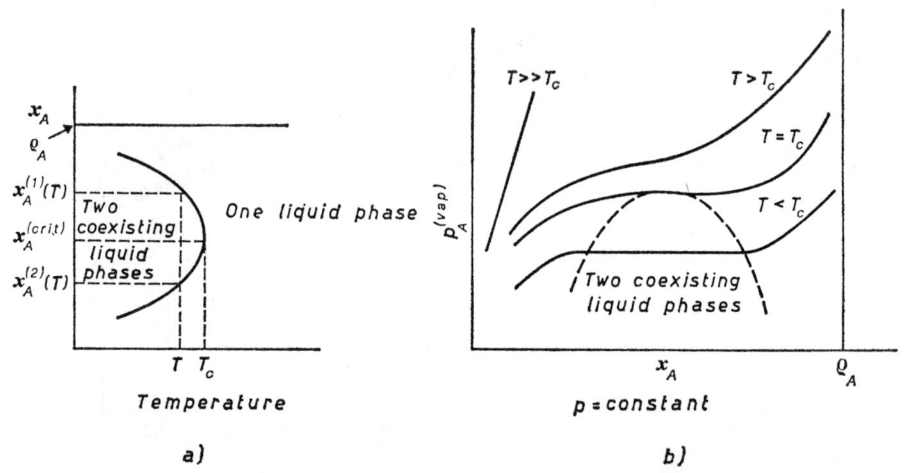

FIGURE 8. Partially miscible binary liquid mixtures: (a) coexistence curve; (b) partial vapour pressure.

above) a consolate (or critical) temperature T_c: below (sometimes above) two phases are present (respectively rich in component A or B) while above (sometimes below) an homogeneous solution is formed. The shape of the solubility curve (Fig. 8a), which gives the mole fraction of the two phases in equilibrium at a temperature smaller than T_c, is similar to the liquid vapour co-existence curve. As a matter of fact the results in many binary systems can be expressed by

$$\frac{x_A^{(1)} - x_A^{(2)}}{\rho_A} = \text{const}\left(1 - \frac{T}{T_c}\right)^{\beta} \tag{6}$$

where $x_A^{(1)}$ and $x_A^{(2)}$, the mole fractions of component A in the two phases, are analogous to the densities in the one component system. The value of β given by experiment is still very near to $\frac{1}{3}$; in the system CCl_4—C_7H_{16}, Thompson and Rice (1964) have found $\beta = 0.33 \pm 0.02$. For a system of this kind if one measures the partial vapour pressure of component A

(p_A^{vap}) above the solution, maintaining constant the external pressure p, one gets isotherms (Fig. 8b) very similar to the isotherms for one component systems.

Moreover, in the critical region, around T_c, critical opalescence is observed, because large composition fluctuations occur and the refractive index depends on composition.

(3) *Ferromagnets.* Let us consider a single domain, the magnetization **M** as well as the applied magnetic field in a z direction coinciding with one

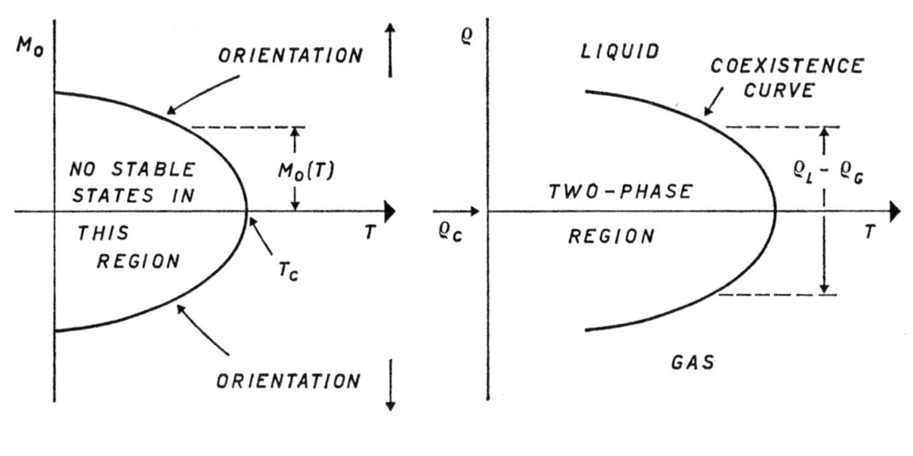

FIGURE 9. Coexistence curves: (a) ferromagnets; (b) simple fluid.

"easy-magnetization" direction. The characteristic feature of these materials is the existence of a spontaneous magnetization

$$M_0(T) = \lim_{H\to 0+} M(H, T) \tag{7}$$

below a particular temperature, T_c, the Curie temperature. In equation (7) H is the "true" field acting on the system, i.e. the external field corrected for demagnetization due to the specimen free poles. Figure 9 shows how the equilibrium magnetization $M_0(T)$ within a single domain (oriented up or down) changes with temperature; in the same figure is presented the analogous co-existence curve for a one-component fluid. In the analogy M and H correspond respectively to ρ and p. The shape of the curve can be represented by

$$\frac{M_0(T)}{M_0(0)} = \text{const}\left(1 - \frac{T}{T_c}\right)^{\beta} \tag{8}$$

with the experimental value $\beta \simeq \frac{1}{3}$.

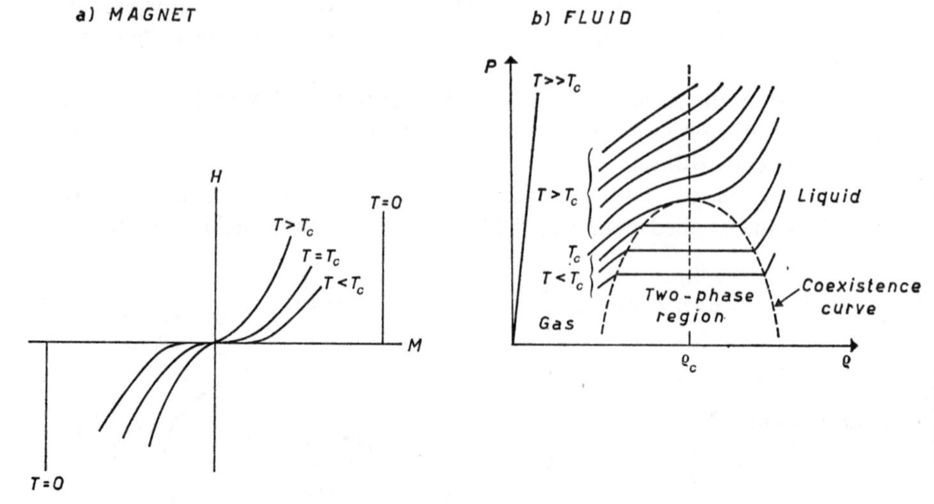

FIGURE 10. Isotherm: (a) H-M in ferromagnets; (b) in simple fluids.

Above T_c the system is in the "paramagnetic region" where M changes continuously with H; below T_c the equilibrium magnetization changes discontinuously as H goes through zero. These properties are shown clearly by means of the isotherms in the plane M, H. Figure 10 gives the shape of these lines, and shows also the similar isotherms in the plane p, ρ for a simple fluid. The shape of the critical isotherm can be fitted by an expression

$$\frac{H}{H_I} = \text{const}\left|\frac{M}{M_0(0)}\right|^{\delta} \tag{9}$$

analogous to the expression for fluids. Here $H_I = \dfrac{k_B T_c}{m}$ and m is the magnetic moment per spin.

The magnetic susceptibility

$$\chi = \left(\frac{\partial M}{\partial H}\right)_T \tag{10}$$

is the analogue of the fluid compressibility (K_T). It diverges as $T \to T_c$ (Fig. 11).

To complete the analogy between ferromagnets and fluid systems Fig. 12 shows the (H, T) phase equilibrium diagram and the analogous p-T diagram for simple fluids (see Fig. 2). The dotted line suggests a way to pass continuously from one phase to the other through states above the critical point; the arrow indicates the predominant spin configuration in the various regions of the H, T diagram.

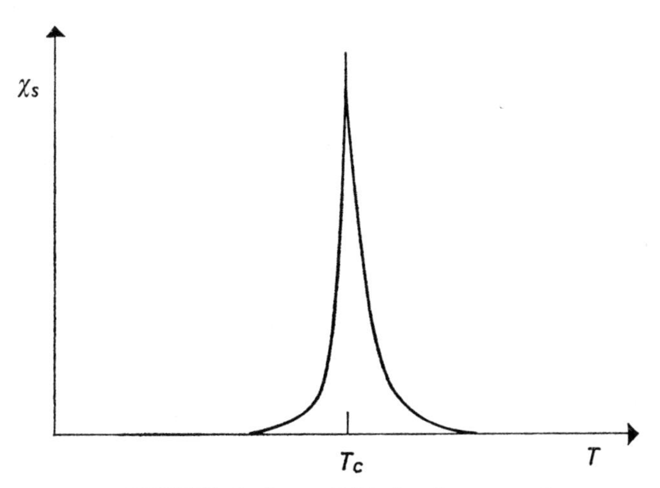

FIGURE 11. Susceptibility in a ferromagnet.

The specific heat contribution due to magnetization (which appears superimposed on the lattice contribution) can be obtained by experiment if T_c is small compared with the Debye temperature. This contribution is of the λ type and roughly shows a weak divergence as T_c is approached. Figure 13 gives the results for the total specific heat in EuS and EuO by Teaney (1966). The anomaly occurs at the Curie point (indicated on the figure). The existence of large fluctuations in the electronic spin distribution near T_c is detected by scattering of neutron beams in analogy to the critical scattering of light in fluids (opalescence).

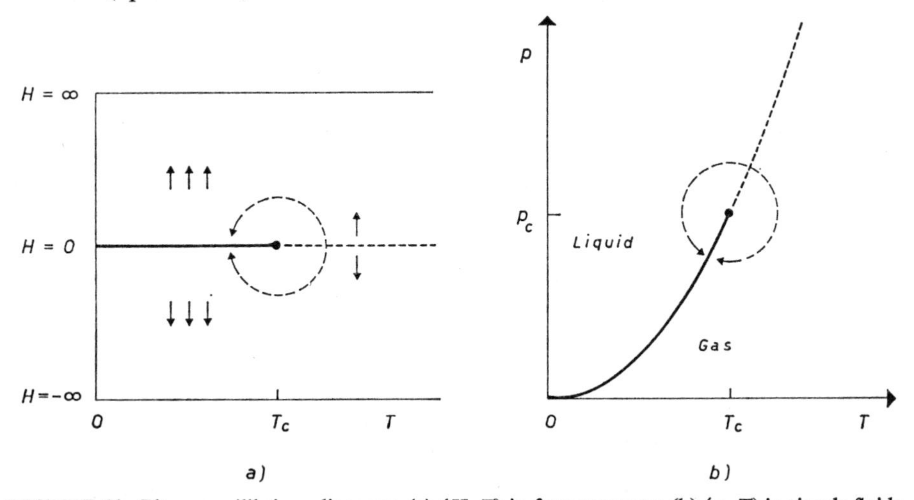

FIGURE 12. Phase equilibrium diagram: (a) (H, T) in ferromagnets; (b) (p, T) in simple fluids.

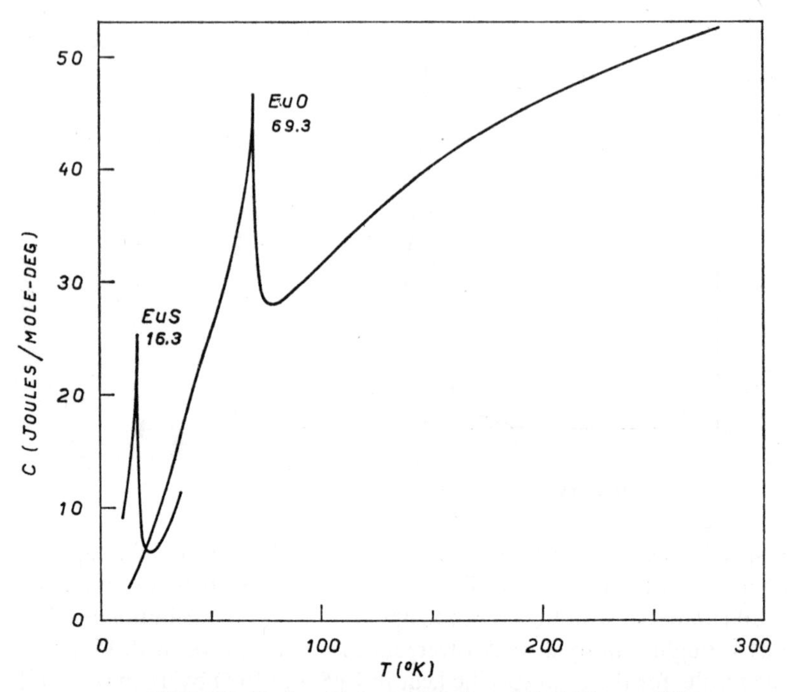

FIGURE 13. Specific heat of EuO and EuS (Teaney 1966).

(4) *Antiferromagnets.* In these materials two sublattices of opposite spins exist at temperatures lower than the Néel temperature T_c. Figure 14a gives a diagram of an idealized one-dimensional antiferromagnet. In this figure arrows indicate the time-averaged position of spins around which the spins fluctuate rapidly. Because the total domain is non-magnetic, the magnetic experiments are performed with a microscopic technique such as nuclear magnetic resonance or neutron diffraction; calorimetric experiments are also made to determine the specific heat dependence upon temperature.

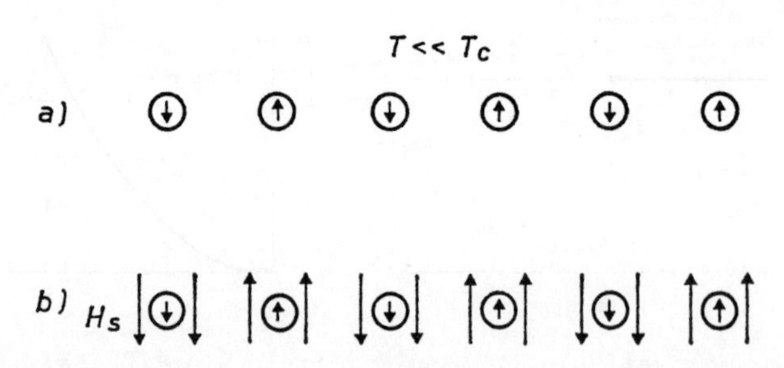

FIGURE 14. Scheme of adjacent spin disposition in a one-dimensional antiferromagnet (a) and schema of idealized experiment for defining the staggered susceptibility (b).

The analysis of results leads to a consideration of the spontaneous magnetization M_s of each sublattice. Its dependence on temperature has been established with great accuracy in some materials; in MnF_2 for instance, Heller and Benedek (1962) by using NMR techniques have made accurate measurements (1 part in 20,000) approaching T_c within a few parts in 10^5. The experiments can be fitted by

$$\frac{M_s(T)}{M_s(0)} = 1.20\left(1 - \frac{T}{T_c}\right)^{1/3}. \tag{11}$$

which again is similar to the co-existence curve of a simple fluid.

To obtain something similar to the isotherms for ferromagnets one should imagine applying to the one dimensional distribution of spins a staggered magnetic field H_s which switches alternately from H to $-H$ passing from one spin site to the neighbour (i.e. applying the field H to each sublattice in the direction of the corresponding spins).

One also defines a staggered susceptibility

$$\chi_s = \left(\frac{\partial M_s}{\partial H_s}\right)_T. \tag{12}$$

The staggered susceptibility evaluated at $H = 0$, has the same behaviour as the susceptibility of a ferromagnetic (Fig. 11). Indications on χ_s can be extracted from the results of neutron diffraction study of magnetic fluctuations. The study of critical fluctuations with NMR techniques has also shown the phenomenon of slowing down, similar to that observed in fluids, by means of light scattering experiments. The results on the magnetic contribution to the specific heat are of the same type as for ferromagnetic materials.

(5) *Ordering crystalline binary alloys.* In some binary alloys each of the two different kinds of atoms are regularly disposed on sublattices at temperatures lower than a critical one. A well studied case is the 50% Cu–Zn alloy (betabrass). The order is very similar to the one existing in antiferromagnets when $T < T_c$ and becomes random above T_c. It is possible to consider here also a quantity P_0 (order parameter) which describes the amount and kind of order in the neighbourhood of the critical point, i.e. analogous to $(\rho - \rho_c)$ in simple fluids and to $M_0(T)$ in ferromagnets. Neutron scattering experiments (Als-Nielsen and Dietrich, 1967) have allowed a very accurate study of critical fluctuations in the beta brass system. The specific heat also shows here a marked λ type anomaly. The dependence of the order parameter on temperature (analogous to the coexistence curve) is

$$P_0 \sim (T_c - T)^{\beta}, \tag{13}$$

with $\beta = 0.305 \pm 0.005$.

(6) *Superfluid helium.* Figure 15 gives the phase diagram of helium, the superfluid phase is entirely enclosed. The transition across the λ line, i.e. between normal fluid and superfluid, seems to be continuous along its whole length and it is of the second order type O. A deeper study shows various analogies between this transition and that occurring in antiferromagnets (at least in small magnetic fields). In the superfluid transition of helium a finite

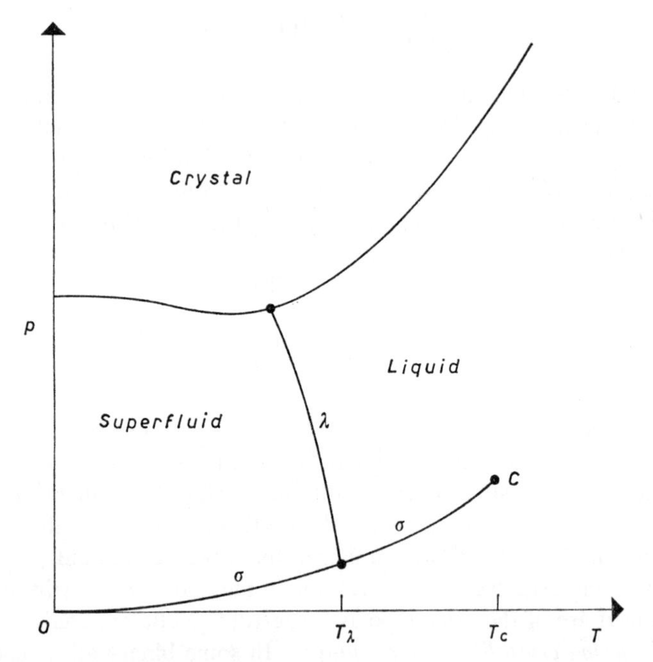

FIGURE 15. Scheme of the phase diagram of ^4He: σ is the vapour pressure line and ends at the critical point C of the gas-liquid system.

fraction of atoms of the system falls into the same quantum state. It is possible to show that in this case the order parameter is the wave function $\langle\psi(r)\rangle$ of this state; it is a complex order parameter, i.e. with two components. The situation is therefore more complicated than in other processes (e.g. the liquid-gas transition); moreover it is not possible to control the thermodynamic quantity which is conjugate to the order parameter (e.g. the magnetic field which is conjugate to M, in magnets). Notwithstanding these differences the superfluid transition is accompanied by critical processes which are in many respects analogous to those of other systems. Experiments by Buckingham *et al.* (1961) on the specific heat along the vapour pressure line (Fig. 16) show that the specific heat has a logarithmic divergence. The results over

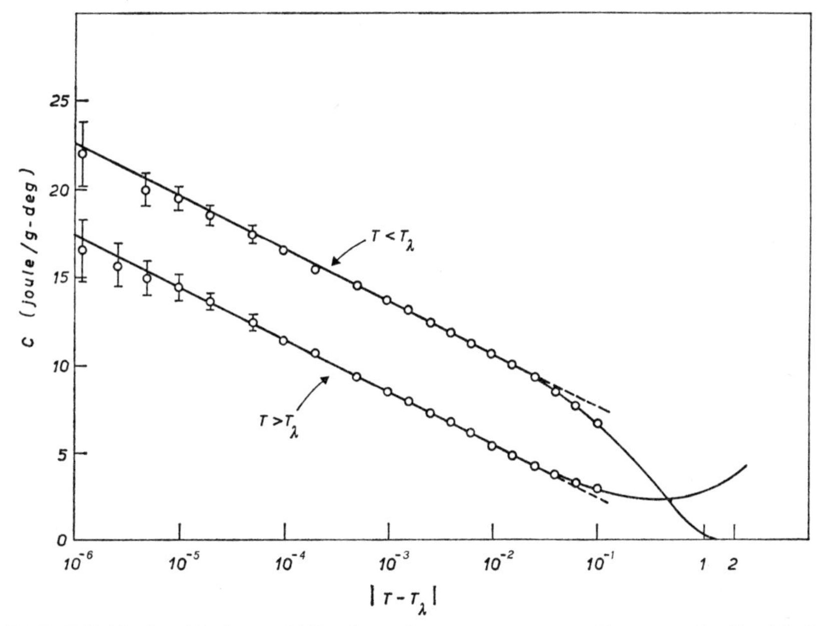

FIGURE 16. Specific heat of ^{4}He along the vapour pressure line near T_λ (Buckingham *et al.* 1961).

about four decades of $|T - T_c|$ can be accounted for by an expression of the type:

$$\frac{c_\sigma}{k_B} = A^+ \ln\left|\frac{T}{T_c} - 1\right| + B^+ \qquad (T \geqslant T_c)$$

$$= A^- \ln\left|\frac{T}{T_c} - 1\right| + B^- \qquad (T \leqslant T_c) \tag{14}$$

with $A^- = A^+$ and $B^- > B^+$.

Moreover, accurate measurements (Clow and Reppy, 1966; Tyson and Douglas, 1966) of the superfluid density ρ_s just below T_λ give

$$\rho_s(T) \sim (T_\lambda - T)^\zeta \tag{15}$$

with $\zeta = 0.666 \pm 0.006$ and because in the Landau-Ginzburg theory ρ_s is proportional to the square of the order parameter

$$\rho_s \sim |\langle\psi\rangle|^2$$

one gets the indication that

$$\langle\psi\rangle \sim (T_\lambda - T)^{0.33}, \tag{16}$$

which is close to the order parameter dependence on $(T - T_c)$ for other transitions.

(7) *Other critical systems.* There are other systems where critical points exist but we mention here only superconductors, ferroelectrics, antiferroelectrics. We do not discuss their characteristics both because much less information is available, and because their special properties require careful discussion which is beyond the scope of this article.

This short review of experimental results has shown the variety of processes in which critical points occur and the many striking similarities in the behaviour of the characteristic equilibrium properties of widely different systems. We shall see that certain conclusions arise from this comparison. In what follows, however, we will refer mainly to fluid and ferromagnetic systems.

B. CLASSICAL THEORIES

The term "classical" is here used to indicate those theories which are based on well established general ideas of thermodynamic systems and which until recently were the only ones developed.

The first attempts to produce an equation of state which could account for critical phenomena were made in the simple fluid case. In 1873, a few years after Andrews experiments in CO_2 (1869), Van der Waals worked out his well known equation

$$\left(p + \frac{a}{v^2}\right)(v - b) = \frac{R}{M} T \tag{17}$$

starting from the ideal gas state equation and taking account of the finite size of molecules and the intermolecular forces. The molecules were assumed to attract each other with (very) long range forces when separated from one another and to repel as hard spheres when in contact; $v = \frac{1}{\rho}$ is the specific volume, M the molecular weight, R the gas constant, a and b phenomenological parameters. Such an equation of state is able qualitatively to account for the shape of isotherms in the gaseous and liquid state as well as for the continuity of the two phases (Fig. 17). It gives however a wrong shape in the vapour-liquid region, but this is no surprise because the equation was developed for the case when only one phase is present. The qualitative agreement of shape between theoretical and experimental isotherms in the two-phase region can however be established by using the Maxwell equal area rule, i.e. by substituting part of the Van der Waals isotherm with a horizontal line in the two phase region such that the two regions in the ρ, p plane between the two lines (Van der Waals and horizontal) have equal area.

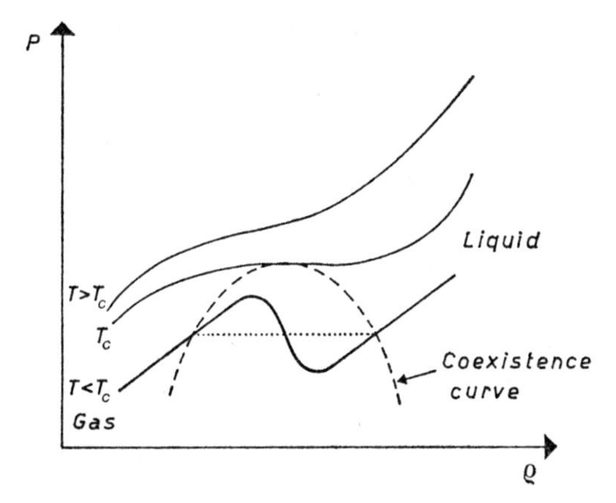

FIGURE 17. Van der Waals isotherm (solid line), coexistence curve (dashed line) and Maxwell equal area rule for determining the isotherm in the two phase region.

This construction can be justified from thermodynamics by the necessity that the points representing co-existent (liquid and gas) phases have the same free energy $G(p, T)$ for equilibrium at constant p and T.

The isotherm at the critical point has an inflection with a horizontal tangent and therefore the position of the critical point is obtained by the conditions

$$\left(\frac{\partial p}{\partial \rho}\right)_T = 0 \qquad \left(\frac{\partial^2 p}{\partial \rho^2}\right)_T = 0. \tag{18}$$

It is possible with this state equation to calculate the quantities of interest: they can be expressed as a power series in $\varepsilon = \dfrac{T - T_c}{T_c}$, so that the leading term determines the behaviour of each quantity in the immediate proximity of the critical point. One gets: (1) the shape of the co-existence curve

$$\Delta\rho \sim |\varepsilon|^{1/2} \tag{19}$$

with $\Delta\rho = \rho_L - \rho_G$:

(2) the shape of the critical isotherm in the proximity of the critical point

$$\Delta p \sim \Delta\rho^3 \tag{20}$$

where

$$\Delta p = \frac{p - p_c}{p_c}:$$

(3) the isothermal compressibility along the critical isochore

$$K_T \sim \varepsilon^{-1}: \tag{21}$$

(4) the specific heat for one kmole along the critical isochore

$$C_v \sim \tfrac{3}{2}R + \tfrac{9}{2}R[1 - \tfrac{28}{25}\varepsilon] \quad \varepsilon < 0$$
$$C_v \sim \tfrac{3}{2}R \qquad \varepsilon > 0. \tag{22}$$

The specific heat has therefore only a *discontinuity* at the critical point: $\Delta C_v = \frac{9}{2}R$.

Similar results have been obtained in ferromagnetic systems. In 1895 P. Weiss' experiments in iron resulted in curves of magnetization vs temperature for different magnetic fields very similar to the isobars (density-temperature) for CO_2 (Fig. 18) and he was led to a state equation equivalent to the Van der Waals equation. This goal was reached by considering attractive forces between elementary magnets in the form of an internal average molecular field proportional to magnetization (M) which is added to the external field to give the total field ($H + \alpha M$) acting on each magnet; α, the Weiss constant, depends on the material. The evaluation of the critical parameters leads to results equivalent to those found for fluids; Table 1 gives this comparison. The specific heat at zero field also corresponds to $M = 0$

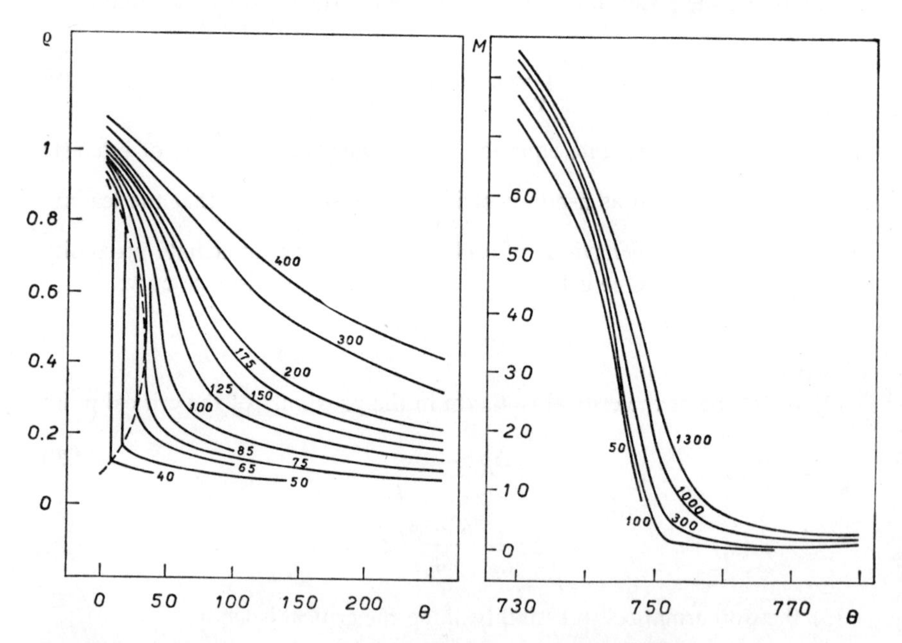

FIGURE 18. Density vs temperature isobars in CO_2 (Amagat) and Magnetization vs temperature curves for iron at different magnetic fields (Weiss 1907).

Table 1
Correspondence of critical quantities in simple fluids and ferromagnets

Fluids		Ferromagnets	
Coexistence curve		Spontaneous magnetization	
$\Delta\rho \sim \lvert\varepsilon\rvert^{1/2}$		$M_0 \sim \lvert\varepsilon\rvert^{1/2}$	
Critical isotherm		Critical isotherm	
$\Delta p \sim \Delta\rho^3$		$H \sim M^3$	
Compressibility along the critical isochore		Susceptibility above T_c	
$K_T \sim \varepsilon^{-1}$		$\chi_0 \sim \varepsilon^{-1}$	
Specific heat along the critical isochore		Specific heat in zero field	
$C_V \sim \frac{3}{2}R + \frac{9}{2}R[1 - \frac{28}{25}\varepsilon]$	$\varepsilon < 0$	$C_{H=0} \sim \frac{5}{2}R[1 - \frac{13}{5}\varepsilon]$	$\varepsilon < 0$
$\sim \frac{3}{2}R$	$\varepsilon > 0$	$C_{H=0} = 0$	$\varepsilon > 0$

(see Fig. 12) as Fisher has shown: for $T > T_c$ in fact $H = 0$ implies $M = 0$; for $T < T_c$, when two phases exist, the $M = 0$ specific heat also corresponds to the $H = 0$ one because only then may two opposite domains exist in the same specimen to produce zero magnetization. The first clarification of this coincidence of results in the critical region deduced by the Van der Waals equation for fluids and by the Weiss treatment of ferromagnets, comes from the fact that the effect of molecular interactions as taken into account by Van der Waals is almost equivalent to assuming that the particles move freely in a "mean field" created by all other particles. Reif (1965) has shown this, assuming for the mean field an average potential as in Fig. 19a. Kac *et al.* (1963) have also shown that the Van der Waals equation follows in the case of a potential distribution as indicated in Fig. 19b (i.e. $U(r) = \infty$, $r < r_0$; $U(r) = ae^{-\kappa r}$, $r > r_0$ with the parameter κ determining the range of the potential in the limit $\kappa \to 0$, when the range of the potential becomes infinite while its value becomes infinitely weak. A more general explanation of the coincidence of critical parameters resulting from the Van der Waals and Weiss treatments stems from the fact that both equations are consistent with

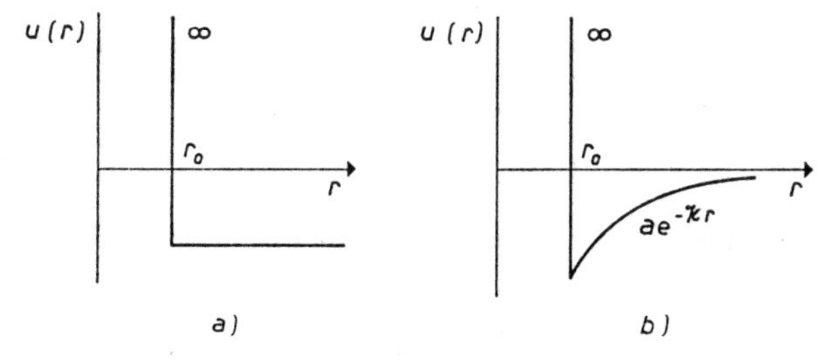

FIGURE 19. Effective potential $U(r)$.

an analytic behaviour of the thermodynamic functions at the critical point. It is indeed possible to show that if one assumes a free energy non-singular at T_c, and expands it as a Taylor power series one is led to the parameters of Table 1, the critical point being defined by means of

$$\left(\frac{\partial p}{\partial \rho}\right)_T = 0 \qquad \left(\frac{\partial^2 p}{\partial \rho^2}\right)_T = 0 \tag{23}$$

or

$$\left(\frac{\partial H}{\partial M}\right)_T = 0 \qquad \left(\frac{\partial^2 H}{\partial M^2}\right)_T = 0. \tag{24}$$

The explanation of critical phenomena in order-disorder transitions in alloys given by Bragg and Williams in 1934 is similar to the Weiss theory. Bragg and Williams however first introduced the concept of a "long range order" which is non-zero below T_c. In 1937 Landau evolved a more general theory in which an "order parameter"† is introduced and the assumption is again made of analyticity of the free energy. The free energy is expanded as a power series in the order parameter with coefficients which are a function of T and p. The assumption is made that a change of symmetry occurs at the critical point. This theory has a very wide application from simple fluids (where the order parameter is $\rho - \rho_c$) and ferromagnets (the order parameter is the magnetization) to superfluids, superconductors etc. The critical behaviour described by this theory still corresponds to the results given in Table 1.

C. SCATTERING

The considerations of state equations and thermodynamical functions of the preceding section gives the classical treatment of the macroscopic behaviour of critical phenomena. Let us examine now the information obtainable from microscopic studies and in particular from light scattering experiments in fluids and from the corresponding classical theory.‡

† The order parameter is introduced for each critical system as a quantity which describes the kind and the amount of order existing in the neighbourhood of the critical point and has the following properties (which are here expressed with reference to systems for which the two phases may exist below a T_c): (a) the order parameter may vanish above the critical point, but it is non-zero just below T_c; (b) it can approach zero continuously as $T \to T_c$ from below; (c) below T_c, the order parameter is not entirely determined by the external conditions, i.e. it can take two or more values under identical values of variables which specify the thermodynamic state.

‡ Light is e.m. radiation in a limited range of wavelengths: it can probe fluctuations over distances equal or larger than $\frac{\lambda}{2}$, i.e. over distances which are still many molecular separations. This allows us to use thermodynamics in the theory. Complementary information can be obtained by X-rays or neutron scattering.

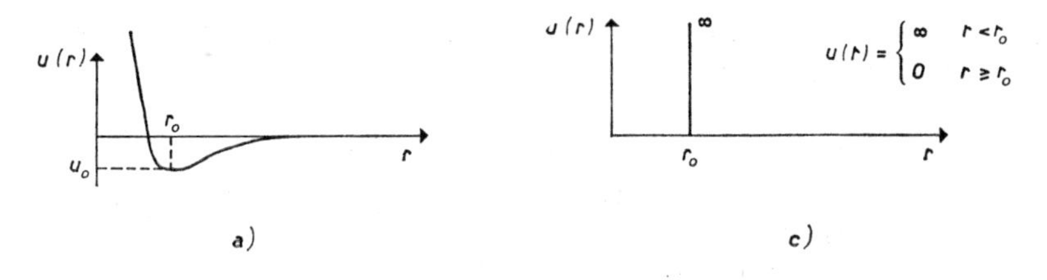

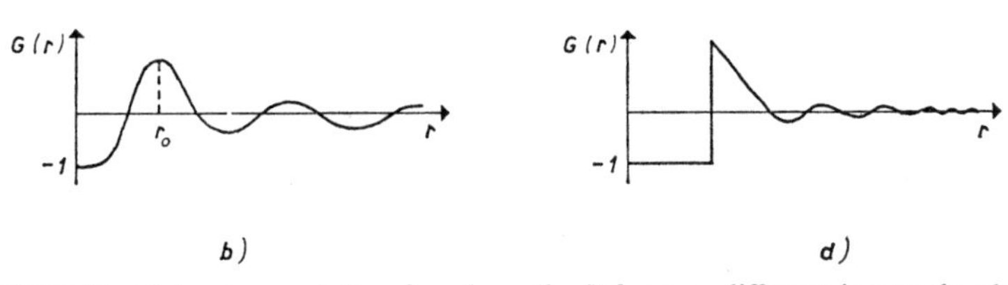

FIGURE 20. Net pair correlation functions (b, d) for two different intermolecular potentials (a and c respectively).

It is important at this point to introduce the *net pair correlation function* (or density-density correlation function)

$$G(\mathbf{r}) = g(\mathbf{r}) - 1 \tag{25}$$

where $g(\mathbf{r})$ is the radial distribution function measuring the probability of finding a molecule at position $\mathbf{r}$ distant from a given molecule. In a one-phase system $G(\mathbf{r}) \to 0$ as $\mathbf{r} \to \infty$. Figure 20 gives the general behaviour of $G(\mathbf{r})$ for two different shapes of intermolecular potential (Lennard-Jones and square well): after a few atomic spacings (~ 20 Å) the function is near zero.

One may also consider the Fourier transform

$$G(\mathbf{k}) = \int \exp(i\mathbf{k} \cdot \mathbf{r}) G(\mathbf{r})\, d\mathbf{r}. \tag{26}$$

The thermodynamic theory of fluctuations (Landau and Lifshitz, 1958) allows us to express the density fluctuations by means of the isothermal compressibility

$$\langle (\Delta\rho)^2 \rangle = \rho^2 k_B T K_T, \tag{27}$$

where $\langle (\Delta\rho)^2 \rangle$ is the average in the equilibrium ensemble taken per unit volume. The isothermal compressibility can in turn be related to the correlation function (fluctuation theorem)

$$k_B T \rho K_T = \frac{K_T}{K_{T,I}} = 1 + \rho \int G(\mathbf{r})\, d\mathbf{r}, \tag{28}$$

$K_{T,I}$ being the compressibility of an ideal gas. The mathematical form of equation (28) shows that the divergence of K_T as $T \rightarrow T_c$ implies a slower and slower decay of $G(\mathbf{r})$, i.e. that the "range" of the net pair correlation function increases continuously as the critical point is approached. When this range ("correlation length") becomes comparable with the light wavelength a strong scattering occurs (*critical opalescence*). Moreover equations (27) and (28) clearly show that the increases in density fluctuations, in compressibility and in the range of the density-density correlation function, as $T \rightarrow T_c$, are intimately connected phenomena.

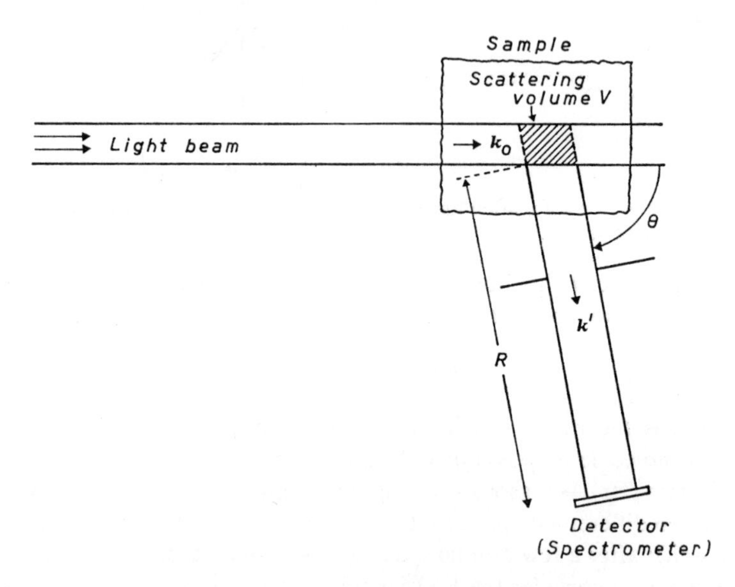

FIGURE 21. Scheme of a scattering experiment.

In a scattering experiment (Fig. 21) a collimated monochromatic light beam† irradiates the system and one observes the light scattered at an angle θ by a volume V and received at a distance R. In our considerations we assume that only single scattering occurs (Born approximation) and that the scattering takes place "quasi-elastically", i.e. that the energy exchanged between radiation and the system is small in comparison with the energy of the incident radiation. This means that the energy of the incident photons is much larger than the typical excitation energy of the system. In these conditions the wave

† At the present time, experiments are conducted by using lasers because of the greater intensity and, in particular, monochromaticity they offer in comparison with thermal sources. In this way it is possible to make not only intensity measurements but also a spectrum determination of the scattered light. We will return to the spectrum determination later.

vectors of the incident radiation $\mathbf{k}_0$, of the scattered radiation $\mathbf{k}'$ and the wave vector $\mathbf{k}$, or momentum transfer, are related by (Fig. 22)

$$\mathbf{k} = \mathbf{k}' - \mathbf{k}_0$$

and

$$|\mathbf{k}| = 2k_0 \sin\frac{\theta}{2} = \frac{4\pi}{\lambda}\sin\frac{\theta}{2} \tag{29}$$

since $k' \simeq k_0$. Here λ is the light wavelength. Let us call $I_0(\mathbf{k})$ the intensity of scattered light which would have been measured if the particles were uncorrelated; it depends only on the properties of isolated particles. When

FIGURE 22. Wave vectors in a scattering experiment.

instead a correlation among particles exists, the scattered waves from individual particles are related by phase factors and interfere; the scattered intensity $I(\mathbf{k})$ contains information on the correlation. It is possible to show, (Fisher 1967), that†

$$\frac{I(\mathbf{k})}{I_0(\mathbf{k})} = 1 + \rho G(\mathbf{k}) \tag{30}$$

Thus the intensity of radiation scattered through a wave vector $\mathbf{k}$ is related to the corresponding Fourier transform of the density-density correlation function. For the forward-scattering ($\theta = 0$, $k = 0$), being $G(K = 0)$ given by the integral of $G(\mathbf{r})$ over all space

$$\frac{I(0)}{I_0(0)} = 1 + \rho\int G(\mathbf{r})\,\mathrm{d}\mathbf{r} \tag{33}$$

† The analysis of scattering experiments is frequently done in terms of appropriate cross sections which depend only on the scatterers and not on the geometry of the experiment or the power of the source. The cross section of interest here is

$$\frac{1}{V}\frac{\mathrm{d}\sigma}{\mathrm{d}\Omega} = \frac{I(\mathbf{k})}{I_0{}^*}\frac{R^2}{V}\ (\mathrm{cm}^{-1}). \tag{31}$$

$I_0{}^*$ being the average incident intensity. This cross section represents the flux of light scattered into a unit solid angle per unit scattering volume per unit incident intensity. It is sometimes called the "Rayleigh ratio". It is found that it can be

$$\frac{1}{V}\frac{\mathrm{d}\sigma}{\mathrm{d}\Omega} = \frac{N}{V}|A|^2\, S(\mathbf{k}) \tag{32}$$

where A is the coherent scattering amplitude of the field scattered by each of the N particles in the volume V and $S(\mathbf{k})$, the *structure factor*, characterizes the interference of the various scattered waves. In the ideal gas limit (uncorrelated scatterers) $S(\mathbf{k}) = 1$.

and from (28)

$$\frac{I(0)}{I_0(0)} = \frac{K_T}{K_{T,I}} \tag{34}$$

The zero-angle scattering intensity therefore diverges as K_T when $T \to T_c$.

The first qualitative explanation of the scattering in the critical region, which constitutes the so called classical theory, is due to Ornstein and Zernicke (1914, 1916) and essentially consists of deriving, by making a number of assumptions, an expression for $G(\mathbf{r})$ of the form

$$G(\mathbf{r}) = A \frac{e^{-\kappa r}}{r} \tag{35}$$

where A is a coefficient which changes slowly with temperature and density and $\kappa = \frac{1}{\xi}$ is usually called the inverse correlation range while ξ is the *coherence* or *correlation length*. Expression (35) gives essentially an exponential decay with r. If one confronts it with the diagram of $G(r)$ in Fig. 20, it is evident that it is at best correct only asymptotically, i.e. (35) represents only the long-range tail of $G(\mathbf{r})$. This is however the part of $G(r)$ which is of interest in critical scattering.

We will not discuss in detail here the Ornstein-Zernicke theory but will limit ourselves to mentioning the assumptions which were made and whose fundamental validity is still to be demonstrated. The main hypothesis concerns the net pair correlation function $G(\mathbf{r})$. Ornstein and Zernicke assume that this function can be divided in two parts: the "direct correlation" between the two molecules considered (whose relative positions are determined by $\mathbf{r}$) and the "indirect correlation" due to the interaction of other molecules. The direct correlation, $C(\mathbf{r})$ is directly related to the interparticle potential and Ornstein and Zernicke assume it to be short-ranged through the critical region. They assume also that the Fourier transform

$$C(\mathbf{k}) = \int \exp(i\mathbf{k} \cdot \mathbf{r}) C(\mathbf{r})\, d\mathbf{r} \tag{36}$$

has a valid Taylor series expansion about $\mathbf{k} = 0$ for all temperatures up to T_c and that it is a slowly varying function of $\mathbf{k}$. This allowed them to retain only the two leading terms in the series expansion.

Inserting equation (35) into equation (28) and neglecting the constant term, one gets:

$$\frac{4\pi A}{\kappa^2} = k_B T K_T \tag{37}$$

which shows that $\xi = \frac{1}{\kappa}$ increases as $\sqrt{K_T}$ when $T \rightarrow T_c$, i.e. it diverges at the critical point. The expression (35) for $G(\mathbf{r})$ leads to the Fourier transform

$$G(\mathbf{k}) = \frac{4\pi A}{\kappa^2}\left[1 + \frac{k^2}{\kappa^2}\right]^{-1} = k_B T K_T \left[1 + \frac{k^2}{\kappa^2}\right]^{-1} \tag{38}$$

and to an expression for the scattered light near the critical point

$$\frac{I(\mathbf{k})}{I_0(\mathbf{k})} = 1 + \rho k_B T K_T [1 + k^2\xi^2]^{-1} \simeq \rho k_B T K_T [1 + k^2\xi^2]^{-1} \tag{39}$$

since K_T diverges at the critical point. Equation (39) shows that (at fixed density and temperature) the scattered intensity as a function of k is Lorentzian with half width $\xi^{-1} = \kappa$. Because $T \rightarrow T_c$, $\xi \rightarrow \infty$, the scattered light becomes more and more concentrated about $k = 0$, i.e. $\theta = 0$, that is in the forward direction (critical opalescence).

To compare the conclusion of the Ornstein-Zernicke theory with experiment it is useful to consider the inverse scattering intensity

$$\left(\frac{I(\mathbf{k})}{I_0(\mathbf{k})}\right)^{-1} = B\frac{1 + k^2\xi^2}{K_T} \tag{40}$$

where $B = \frac{1}{\rho k_B T}$ can be considered a constant in a small range of temperatures near the critical point. A plot of the inverse scattering intensity versus k^2 (i.e. θ^2) should be a straight line with slope proportional to $\frac{\xi^2}{K_T}$ and intercept (at $\theta = 0$) $\frac{B}{K_T}$. The lines at various temperatures (always near T_c) should be parallel while the intercept should tend to zero as $T \rightarrow T_c$.

Figure 23 gives the result of Thomas and Schmidt (1963) in argon and Fig. 24 gives the results of Kao and Chu (1969) on scattering in the critical region of the binary mixture of n-dodecane and $\beta - \beta'$ dichloro ethyl ether. The experiments allow the determination of $K_T(T)$ and of $\xi(T)$. Figure 25 (Cummins) collects the results for $\xi(T)$ obtained on many simple fluid and binary liquid mixtures with this technique (symbols containing a vertical line), and with a different technique which we shall discuss later (Section II.E).

According to the theory, remembering that the classical relation for the compressibility is

$$K_T \sim |T - T_c|^{-1}$$

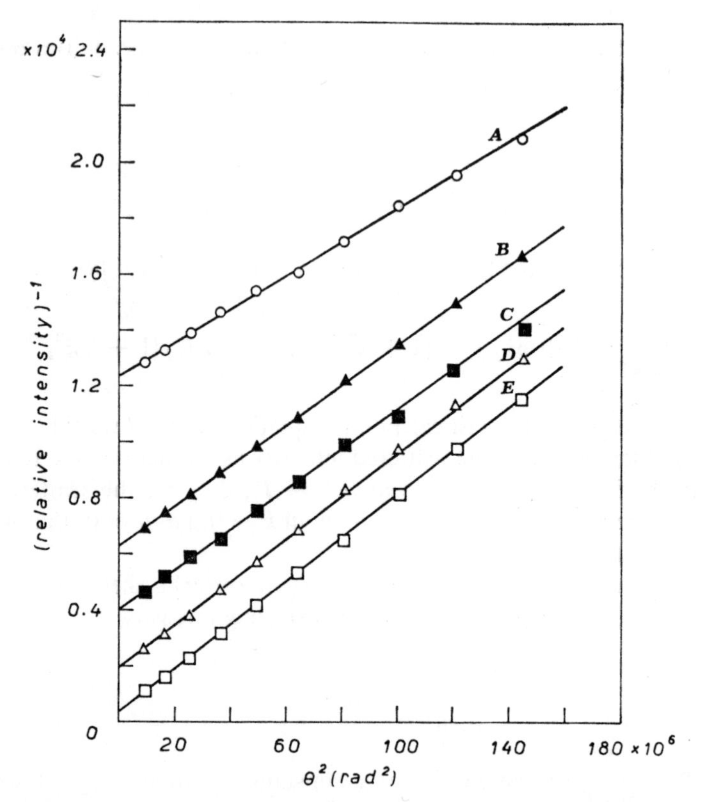

FIGURE 23. Critical scattering in argon for various temperatures on an isobar close to the critical ($p \simeq p_c$): A, $(T - T_c) = 2°K$; B, $(T - T_c) = 1°K$ C, $(T - T_c) = 0.45°K$; D, $(T - T_c) = 0.25°K$; E, $(T - T_c) = 0.05°K$.

the correlation length should vary with $(T - T_c)$ as

$$\xi(T) = \xi_0 \varepsilon^{-1/2}. \tag{41}$$

This expression was used by Ornstein and Zernicke in their original work. Debye (1959) has used a similar expression

$$\xi(T) = \frac{l}{\sqrt{6}} \varepsilon^{-1/2} \tag{42}$$

l^2 being the second moment of the intermolecular potential.

D. INADEQUACY OF CLASSICAL THEORIES. POWER LAWS

Classical theories, although able qualitatively to account for the occurrence of critical phenomena, are unable quantitatively to explain the experimental

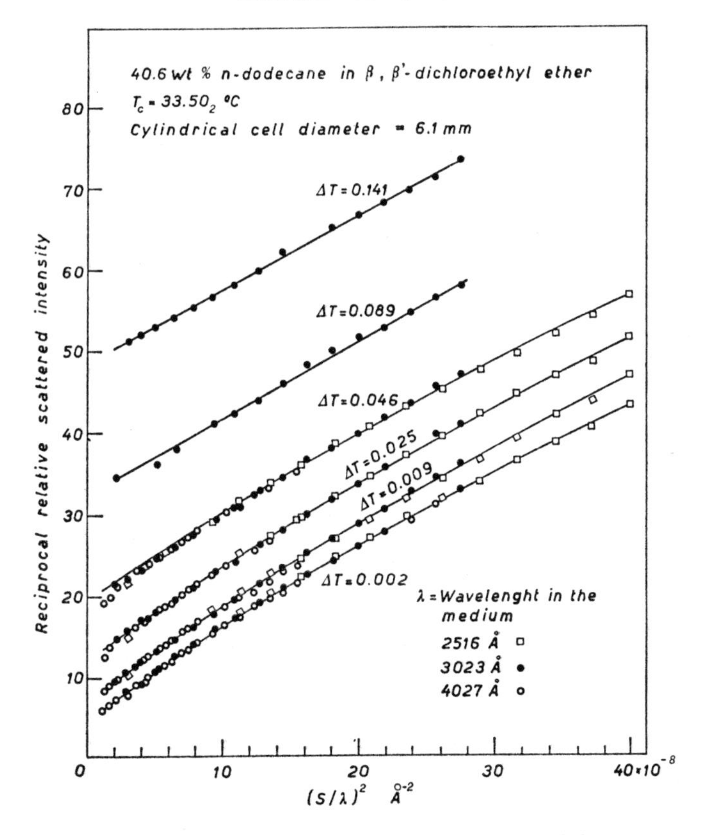

FIGURE 24. Critical scattering for the binary system *n*-dodecane-β,β'-dichloroethylether.

results, this is clearly shown by a comparison of results given in the previous sections. With reference to a fluid, for instance, the classical shape of the coexistence curve should be parabolic while experiment indicates a cubic law; the constant volume specific heat should show only a discontinuity at T_c according to the theory and instead it shows a divergence. In so far as the scattering experiments are concerned the Ornstein-Zernicke expression for the correlation function has been fairly successful in explaining experimental results: deviations however have been detected and Fisher (1964) has suggested a modification of it. Also the dependence of $\xi(T)$ on ε given by experiment deviates slightly from equation (41).

The analytic dependence of quantities obtained through the experiment, on $(T - T_c)$ or on the more commonly used quantity $\varepsilon = \dfrac{T - T_c}{T_c}$, differ, therefore, from the classical predictions. Since the usual case concerns

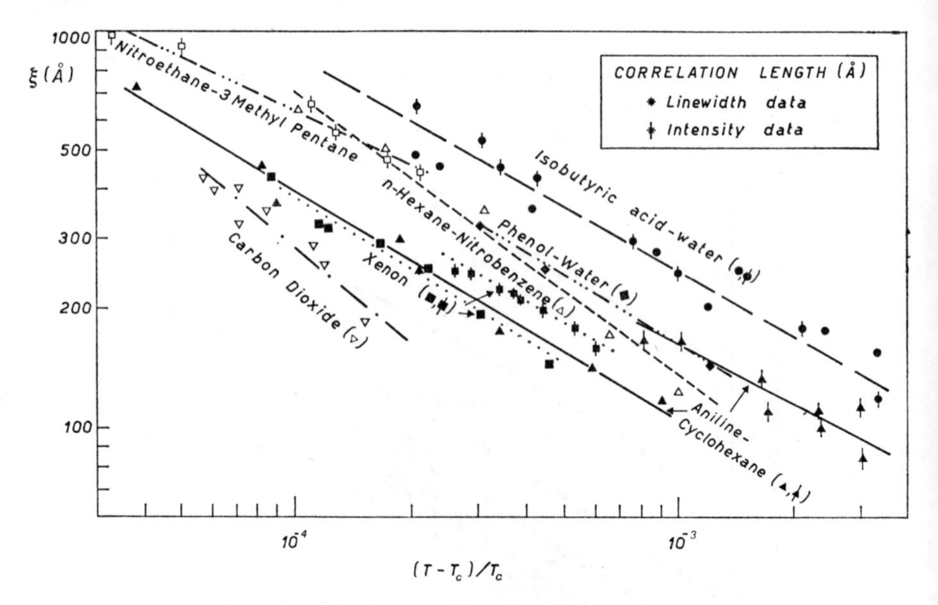

FIGURE 25. Correlation length vs $\varepsilon = \dfrac{T - T_c}{T_c}$ for various fluids and binary liquid mixtures from intensity and linewidth data (Cummins 1970).

quantities which diverge to infinity or converge to zero as $\varepsilon \to 0$, it is usual (following Guggenheim's original suggestion) to assume a power law to be valid, to a first approximation, for each of these quantities X:

$$X \sim |\varepsilon|^\lambda \qquad X > 0 \qquad \text{when} \quad T \to T_c \tag{43}$$

where

$$\lambda = \lim_{\varepsilon \to 0} \left| \frac{\log X}{\log |\varepsilon|} \right|. \tag{44}$$

Of course different values of λ may be necessary for $\varepsilon > 0$, (λ), and for $\varepsilon < 0$, (λ'), and different values of λ' may exist for the different phases (as liquid, gas) below T_c. The special case $\lambda = 0$ may be associated with a pure logarithmic divergence.†

$$\dagger \lim_{\varepsilon \to 0} \frac{\log |\log|\varepsilon||}{\log |\varepsilon|} = \lim_{\varepsilon \to 0} \frac{\dfrac{1}{|\varepsilon|} \dfrac{1}{\log |\varepsilon|}}{\dfrac{1}{|\varepsilon|}} = \lim_{\varepsilon \to 0} \frac{1}{\log |\varepsilon|} = 0 \tag{45}$$

The value $\lambda = 0$ may also correspond either to a finite discontinuity at T_c, or to a regular continuous behaviour through T_c.

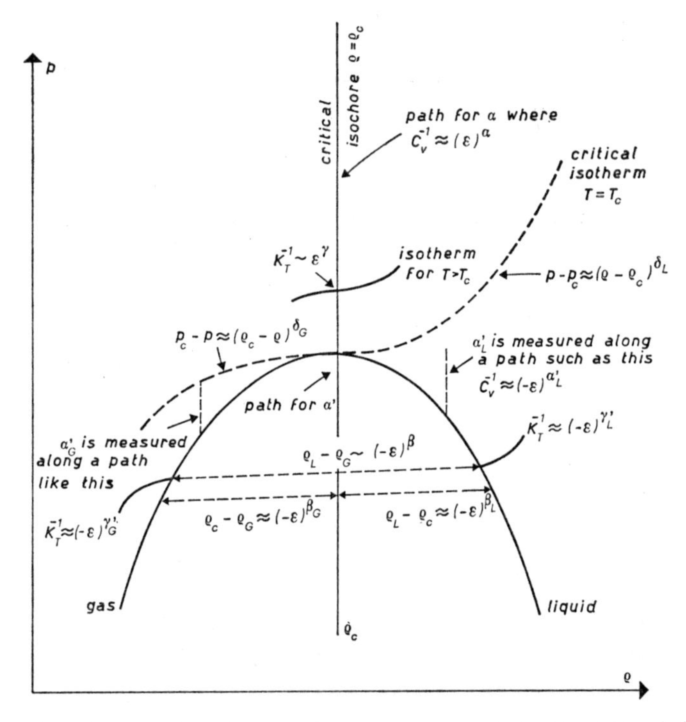

FIGURE 26. Power laws for a single component fluid (after W. C. Marshall).

The exponents λ for the various important quantities have received individual symbols which we introduce here with reference to the simple fluid case with the aid of Fig. 26. Each of them is used also for the corresponding quantities in other critical systems.

(1) Coexistence curve ($\varepsilon < 0$)

$$\rho_L - \rho_G \sim (-\varepsilon)^{\beta} \tag{46}$$

$$\rho_L - \rho_c \sim (-\varepsilon)^{\beta_L} \tag{47}$$

$$\rho_c - \rho_G \sim (-\varepsilon)^{\beta_G} \tag{48}$$

(2) isothermal compressibility along the critical isochore

$$\frac{1}{K_T} \sim \varepsilon^{\gamma} \qquad \varepsilon > 0 \tag{49}$$

$$\frac{1}{K_T} \sim (-\varepsilon)^{\gamma'} \qquad \varepsilon < 0 \tag{50}$$

(3) critical isotherm

$$|p - p_c| \sim |\rho - \rho_c|^{\delta} \tag{51}$$

(4) surface tension

$$M \sim |\varepsilon|^{\mu} \tag{52}$$

(5) specific heat at constant volume along the critical isochore

$$\frac{1}{C_v} \sim \varepsilon^{\alpha} \qquad \varepsilon > 0 \tag{53}$$

$$\frac{1}{C_v} \sim (-\varepsilon)^{\alpha'} \qquad \varepsilon < 0 \tag{54}$$

(6) range of density fluctuations ($\rho = \rho_c$)

$$\xi \sim \varepsilon^{\nu} \qquad \varepsilon > 0 \tag{55}$$

$$\xi \sim -\varepsilon^{\nu'} \qquad \varepsilon < 0 \tag{56}$$

For the net pair correlation function Fisher has proposed a modification of the Ornstein-Zernicke expression (35), and a new critical point exponent, η (not referring to a power of ε), is considered

$$G(\mathbf{r}) \sim \frac{\exp(-\kappa r)}{r^{1+\eta}} \, . \tag{57}$$

Really equation (57) refers to a 3-dimensional system; a more general expression for the large distance asymptotic critical correlation function is

$$G(\mathbf{r}) \sim \frac{1}{r^{d-2+\eta}} \qquad (r \rightarrow \infty) \tag{58}$$

with d the dimensionality.

The predictions of the classical theory correspond to laws of the type (46)–(56) with particular values of exponents:

$$\beta_{\text{class}} = \tfrac{1}{2}; \quad \gamma_{\text{class}} = \gamma'_{\text{class}} = 1; \quad \delta_{\text{class}} = 3; \quad \nu_{\text{class}} = \tfrac{1}{2}; \quad \mu_{\text{class}} = \tfrac{3}{2}$$

The classical value of η is zero. Other exponents can be introduced, as for instance those (gap exponents, Δ_i) which characterize the magnetic field derivatives of the Gibbs potential $G(T, H)$ of a magnetic system.

Table 2 collects the definitions of the most important critical exponents for simple fluids and ferromagnets.

Table 2
Critical exponents: definition in simple fluids and ferromagnets

Critical Exponent	Simple Fluids Definition		Simple Fluids Quantity	Ferromagnets Definition		Ferromagnets Quantity
α	$C_V \sim \varepsilon^{-\alpha}$	$\varepsilon > 0$	Specific heat along critical isochore	$C_{H=0} \sim \varepsilon^{-\alpha}$	$\varepsilon > 0$	specific heat at $H = 0$ which is also C_M (See Fisher, 1967)
α'	$C_V \sim (-\varepsilon)^{-\alpha'}$	$\varepsilon < 0$		$C_{H=0} \sim (-\varepsilon)^{-\alpha'}$	$\varepsilon < 0$	
β	$\rho_L - \rho_G \sim (-\varepsilon)^{\beta}$		liquid gas density difference; coexistence curve	$M_0 \sim (-\varepsilon)^{\beta}$		zero field magnetization
γ	$K_T \sim \varepsilon^{-\gamma}$	$\varepsilon > 0$	isothermal compressibility	$\chi_T \sim \varepsilon^{-\gamma}$	$\varepsilon > 0$	zero field isothermal susceptibility
γ'	$K_T \sim (-\varepsilon)^{-\gamma'}$	$\varepsilon < 0$		$\chi_T \sim (-\varepsilon)^{-\gamma'}$	$\varepsilon < 0$	
δ	$p - p_c \sim \lvert\rho - \rho_c\rvert^{\delta}$		critical isotherm	$H \sim \lvert M\rvert^{\delta}$		critical isotherm
ν	$\xi \sim \varepsilon^{-\nu}$	$\varepsilon > 0$	correlation length	$\xi \sim \varepsilon^{-\nu}$	$\varepsilon > 0$	correlation length
ν'	$\xi \sim (-\varepsilon)^{-\nu'}$	$\varepsilon < 0$		$\xi \sim (-\varepsilon)^{-\nu'}$	$\varepsilon < 0$	
η	$G(r) \sim \lvert r\rvert^{-(d-2+\eta)}$		net pair correlation function; d = dimensionality	$G(r) \sim \lvert r\rvert^{-(d-2+\eta)}$		net pair correlation length; d = dimensionality

The values of exponents are to be found through experiment and the effort of researchers is directed towards more and more accurate results in order to allow a more precise determination of the exponents. We limit ourselves here to the results of the analysis of experimental data of Egelstaff and Ring (1968) and of Kadanoff *et al.* (1967). In Table 3 are shown for a number of selected exponents: (1) the classical values; (2) the best values estimated by Kadanoff *et al.*, from experimental results in simple nonconducting fluids; (3) the values for metals and binary mixtures; (4) the values for ferromagnets; here some experimental results (Ni) which gave β in the proximity of 0.5 are not included; (5) the values given by some theoretical models which we shall discuss later. The comparison of classical and experimental values of exponents shows clearly the inadequacy of classical theories as has already been pointed out. It also shows the existence of close *similarities* among the values found for each exponent. There are however some significant differences among different types of critical phenomena as well as among different systems undergoing the same kind of phase transition. Apart from the case of Ni, already mentioned, many measurements on ferromagnets have indicated for β a value very close to $\frac{1}{3}$ and recent very accurate experiments (Ho and Lister, 1969) have found $\beta = 0.368 \pm 0.005$ for the insulating ferromagnet $CrBr_3$. We observe also the rather large differences between non-conducting and conducting one-component fluids, the latter result pointing to an important role played by the short-range forces.

Classical treatments, as we have already mentioned, have been successful in showing that the critical phenomena are co-operative effects and in reaching the correct conclusions concerning the physical origin of singularities in the equilibrium properties (thermodynamic derivatives). That is that these stem from the large scale fluctuations of the order parameter which are established near the critical points as a consequence of the low cost in free energy that they require.

One may ask what the fundamental reason is for the inability of classical theory to describe quantitatively critical processes (i.e. to furnish exponents which agree with experiments). Following Uhlenbeck (1966) we may say that while classical theory considers forces among particles which are of *very long range*,† the forces responsible for the critical process seem to be of *short range*, i.e. each particle seems to be influenced directly only by a few shells of neighbouring particles. This conclusion is suggested in part by the experimental results already quoted and in part by the indications of theoretical models and calculations which have been developed over a long period in a search for a satisfactory explanation of critical processes. All these theories try to calculate values of exponents and to find in this way which features of interparticle interactions are important in determining the critical behaviour.

† In a fluid a particle moves in a mean field due to all other particles.

E. MODELS AND APPROXIMATE CALCULATIONS OF EXPONENTS

The first statistical atomic model to show a critical point is the two-dimensional Ising model (Onsager, 1944) for a ferromagnet. According to this model, particles are arranged in a regular lattice and each particle (atom in a ferromagnet) can be found in either of two states (spin up or down, parallel to a z direction); moreover, the interactions considered are only those between nearest neighbour particles ($\pm J$ for parallel or antiparallel spin orientation). This model is therefore characterized by *short-range forces.* Ising made the calculation on the one-dimension model ($d = 1$) and no critical point was found. Onsager carried out an exact calculation in a two-dimensional ($d = 2$) crystal in a *zero magnetic field.* He did not establish a complete state equation, which still remains to be done, but was able to calculate the important quantities in zero magnetic field. The more important results are the existence of a critical point and a logarithmic divergence for C_v. The last result was in striking contrast with the indications of the classical theories. The critical exponents furnished by the (exact) calculation for the 2-d Ising model are given in Table 3. Another important conclusion which came from successive studies made by many investigators on various 2-dimensional lattices for the Ising model is that the critical exponents are essentially independent of the lattice structure. Figure 27 gives the results on magnetization for three lattices—quadratic, triangular and honeycomb.

The next obvious step is to pass from the non-physical 2-d Ising model to the 3-d case. Unfortunately however the exact calculational procedure used in two dimensions cannot be extended to 3-d models and one has to use approximate methods. This problem has been the subject of very many investigations and we refer to other reviews (Domb, 1960; Fisher, 1967; Stanley, 1971) for a complete account of the different lines which have been pursued to get reliable answers for the critical behaviour of 3-d Ising models. The most successful procedure has been to consider series expansions of the thermodynamic function of interest in ascending powers either of T ("low temperature expansion") or of $\frac{1}{T}$ ("high temperature expansion"). These expansions are made not about the critical point but about the zero value of T or of $\frac{1}{T}$ so that the critical point is a singularity on or beyond the convergence circle of the expansion and the critical behaviour is given by the asymptotic form of the power series coefficients. The handling of these series to obtain useful results is a rather complex matter: there are non-rigorous "proofs" that the series are really convergent. It is necessary to consider

Table 3
Theoretical and experimental values of critical exponents

Exponent	$T \leqslant T_c$ α'	β	γ'	μ
Classical	0	$\frac{1}{2}$	1	$\frac{3}{2}$
Nonconducting liquid-gas	0.12 ± 0.12	0.346 ± 0.01	1.0 ± 0.3	1.26
Metals (liquid-gas)		0.45		1
Binary mixture		0.33		
Ferromagnetic substances	< 0.16	0.33 ± 0.03		
2-dimensional Ising model	0	$\frac{1}{8}$	$\frac{7}{4}$	
3-dimensional Ising model: lattice gas	$\sim \frac{1}{8}$	$\sim \frac{5}{16}$	$\sim \frac{5}{4}$	

Exponent	T_c δ	$T \geqslant T_c$ α	γ	ν
Classical	3	0	1	$\frac{1}{2}$
Nonconducting liquid-gas	4.4 ± 0.4	0.2 ± 0.2	1.37 ± 0.2	0.64
Metals (liquid-gas)	3.5			0.5
Binary mixture			1	0.5
Ferromagnetic substances	4.1 ± 0.1	< 0.16	1.33 ± 0.03	0.65 ± 0.03
2-dimensional Ising model	15	0	$\frac{7}{4}$	1
3-dimensional Ising model: lattice gas	~ 5	$\sim \frac{1}{8}$	$\sim \frac{5}{4}$	~ 0.638

many terms to have a reasonable hope of finding a good estimate of the asymptotic behaviour because there is no way of estimating the remainder which is neglected. Successive numerical approximation methods have been developed by using graph theory and computers. Confidence in the results furnished by these methods comes from the fact that when applied to 2-*d* models, they give predictions very close to the exact solutions. One has however to always keep in mind that the results are not exact and could be refined by the consideration of further terms in the series. The same results as for the Ising model of a ferromagnet with spin $\frac{1}{2}$ are obtained for a fluid model, the so called *lattice-gas model*, where particles may exist only in

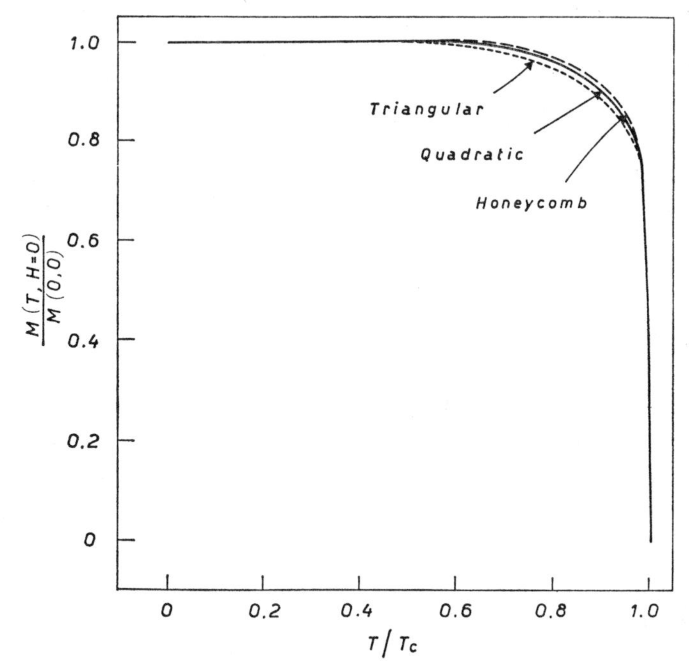

FIGURE 27. Exact calculation of the spontaneous magnetization for different two-dimensional lattices.

correspondence to sites of a regular space lattice. The repulsive forces between molecules ensure that only one molecule can occupy a site, while an attractive force acts between nearest-neighbour sites when occupied. Such a model has, among other characteristics, the advantage of being easily extended to quantum systems.

The values of critical exponents obtained with the approximate calculation in a 3-*d* Ising model (and the gas-lattice model) are given in Table 3. It is evident that they are closer to the experimental values than the classical one; a divergence for C_v is indicated and the value β, close to 0.31, is to be compared with the experimental results of 0.346 in simple fluids and 0.33 in fluid mixtures and ferromagnets (the classical value is 0.5). The results of calculations in 2-*d* and 3-*d* Ising models for various lattice structures (Fig. 27) lead to the suggestion that dimensionality and symmetry are of importance in determining the critical behaviour, while the details of lattice structure have only a slight influence on it.

F. RELATIONS AMONG CRITICAL EXPONENTS

In preceding sections it has been shown how experiments suggest the use of power laws to characterize the behaviour of many physical quantities in the critical region and how models can be devised to guide the calculation of approximate values of exponents by means of series expansions. These expansions have to be considered independently one for each quantity. It is however sound to assume that connections exist among quantities in the critical region so that the next step on the way to finding the fundamental basis of critical phenomena is to investigate possible relations among critical exponents. Essam and Fisher (1963) proposed, on the basis of an empirical model, a relation

$$\alpha' + 2\beta + \gamma' = 0 \tag{58}$$

and Rushbrooke (1963) in the same year established, on thermodynamic grounds valid for all systems, the inequality

$$\alpha' + 2\beta + \gamma' \geqslant 2 \tag{59}$$

Since then many other relations have been proposed and the relations rigorously established are inequalities. We will indicate some of them. We start by presenting Rushbrooke's calculation of his relation in order to indicate one type of argument which has been used to reach interesting conclusions. Let $A(T, M)$ be the Helmholtz free energy for a magnetic system and therefore

$$S = -\left(\frac{\partial A}{\partial T}\right)_M \qquad H = -\left(\frac{\partial A}{\partial M}\right)_T . \tag{60}$$

The thermodynamic stability conditions expressing the specific heat

$$C_M = T\left(\frac{\partial S}{\partial T}\right)_M = -T\left(\frac{\partial^2 A}{\partial T^2}\right)_M \geqslant 0 \tag{61}$$

and the susceptibility

$$\frac{1}{\chi} = \left(\frac{\partial H}{\partial M}\right)_T = \left(\frac{\partial^2 A}{\partial M^2}\right)_T \geqslant 0 \tag{62}$$

may be combined with the relation between specific heats at constant M and constant H

$$C_H = C_M + T\,\frac{\left(\dfrac{\partial M}{\partial T}\right)_H^2}{\left(\dfrac{\partial M}{\partial H}\right)_T} = C_M + T\left(\frac{\partial M}{\partial T}\right)_H^2 \frac{1}{\chi} . \tag{63}$$

Within the range $T \leqslant T_c$ (i.e. $\varepsilon < 0$) and considering the limit $H \to 0$, M becomes the spontaneous magnetization (M_0) and χ the zero field magnetization (χ_0). One then obtains

$$C_{H=0} \geqslant T\left(\frac{\mathrm{d}M_0}{\mathrm{d}T}\right)^2 \chi_0^{-1}. \tag{64}$$

Using the definition of critical exponents

$$C_{H=0} \sim (-\varepsilon)^{-\alpha'}; \qquad \chi_T \sim (-\varepsilon)^{-\gamma'}; \qquad \frac{\mathrm{d}M_0}{\mathrm{d}T} \sim (-\varepsilon)^{\beta-1} \tag{65}$$

and taking the logarithm of equation (64) and dividing by $\log(-\varepsilon)$ the result is obtained

$$-\alpha' \leqslant 2(\beta - 1) + \gamma' \tag{66}$$

which can be put in the form (59). The validity has been extended to fluid systems by Fisher (1964). A second inequality has been derived by Griffiths (1965) from the convexity properties of the free energy

$$\alpha' + \beta(\delta - 1) \geqslant 2. \tag{67}$$

With reference to a one component fluid, Griffiths (1965) has also established the inequality

$$\alpha' + \beta_G \geqslant 0. \tag{68}$$

Griffiths (1965) has also proved the validity of many other relations. From his list we quote

$$\gamma' \geqslant \beta(\delta - 1), \tag{69}$$

$$\gamma(\delta + 1) \geqslant (2 - \alpha)(\delta - 1). \tag{70}$$

Stell (1968) and Buckingham and Gunton (1969) have proved the two inequalities

$$\mathrm{d}\,\frac{\delta - 1}{\delta + 1} \geqslant 2 - \eta \tag{71}$$

$$\frac{\mathrm{d}\gamma'}{2\beta + \gamma'} \geqslant 2 - \eta \tag{72}$$

by using the assumptions of the positivity of the correlation functions, and of their monotonic variation with magnetic field and with temperature.

Fisher (1969) has also proved

$$(2 - \eta)\nu \geqslant \gamma \tag{73}$$

$$d\alpha' > 2 - \eta_E \tag{74}$$

where η_E characterizes the decay of the energy-energy correlation function at $T = T_c$.

Josephson (1967) has instead proved, with some special assumptions

$$d\nu \geqslant 2 - \alpha \tag{75}$$

$$d\nu' \geqslant 2 - \alpha' \tag{76}$$

Under particular assumptions whose validity is difficult to establish and in particular using phenomenological arguments, some equalities have also been suggested, for example equation (58). Essam and Fisher (1963) have also found, for a gas lattice fluid model

$$\gamma = \gamma' \tag{77}$$

$$\alpha = \alpha' \tag{78}$$

Widom (1965), by considering the surface tension in the Van der Waals theory has obtained

$$2\beta + \gamma' = \mu + \gamma \tag{79}$$

and, postulating a particular state equation

$$\gamma' = \beta(\delta - 1) \tag{80}$$

which corresponds to making equation (69) valid as an equality. A large amount of theoretical and experimental work in these last years has been devoted to establishing these inequalities, to comparing them with the predictions of various model calculations and to improving experiments to give more accurate results. The latter may make it possible to determine how well the relations are satisfied as equalities in reality or, more important, what confidence can be placed in the many cases in which they seem to be satisfied as equalities. For instance, if we take the Rushbrooke relation (59) and use the value of exponents of Table 3, it is easy to check that it is valid as an equality for classical models, and for 2-*d* Ising models. The approximate values calculated for 3-*d* Ising model suggest that the same conclusion may be valid there. In the few cases where experimental values of α', β, γ' have been determined the expression $\alpha' + 2\beta + \gamma'$ fails to equal 2 unless the errors are taken into account. The Griffith relation (67)

$$\alpha' + \beta(\delta + 1) \geqslant 2$$

holds as an equality for classical theories and for the 2-*d* Ising model. It can be compared with experiment in He^4. By using the results of Roach and Douglas (1967) and Roach (1968) for β (0.354 ± 0.10) and δ ($3.8 \leqslant \delta \leqslant 4.1$) and those of Moldover (1969) for α' (~ 0.15), one gets for

$$\alpha' + \beta(\delta + 1)$$

the maximum value of 2.03.

Other examples can be given which show the importance of accurate experiments and the difficulty of establishing the validity of the relations as equalities.

These efforts, however, are of great importance in finding out how many independent critical exponents are really needed for the description of a critical system and to give some clues as to the physical quantities of primary importance in critical processes.

G. THE HOMOGENEOUS FUNCTION APPROACH AND SCALING LAWS

In recent years (since 1965) various authors have tried to overcome the difficulties found in developing a coherent treatment of critical phenomena, by making drastic, and usually unproved, assumptions which permit the calculation of a set of exponents and lead eventually to a state equation which can be compared with experiment and the predictions of various models.

This approach has been intriguingly successful in reaching a deeper insight into the nature of critical phenomena.

In 1965 at least two such attempts were made starting from different points but always introducing an unjustified hypothesis which can be reduced to the assumption that the thermodynamic potential of interest has the form of a homogeneous function. Widom (1965) started from an attempt to generalize Van der Waals' state equation to accommodate nonclassical exponents. Domb and Hunter (1965), considering the 2-*d* and 3-*d* models of a ferromagnet, suggested some assumptions on the high order derivatives of M with respect to H which allows an expression for H on the high temperature side of T_c in the form of a series expansion

$$H \simeq b_1 M \varepsilon^{\gamma} + b_3 M^3 \varepsilon^{3\gamma - 2\Delta} + b_5 M^5 \varepsilon^{5\gamma - 4\Delta} + \cdots$$
$$= \varepsilon^{\Delta} F(M \varepsilon^{-\beta}) \tag{81}$$

with $\Delta = \beta + \gamma$ and F a function which may differ from one model to another but is expected to behave smoothly and may even be analytic.

An analogous treatment, starting from some assumptions on correlations in the critical region, was given in 1966 by Patashinskii and Pokrovskii (1966); while Kadanoff, also in 1966, presented a scheme which tries to suggest a microscopic behaviour and which justified better the term "scaling."

We will consider in the present section in general terms the homogeneous function approach and leave to the following one Kadanoff's heuristic introduction of "scaling laws."

Let us start from one thermodynamic potential, the Gibbs function,† for a magnetic system and consider only the singular part near a critical point which we write $G(\varepsilon, H)$. The assumption, which we call the "scaling hypothesis," is now made that $G(\varepsilon, H)$ is a generalized homogeneous function.‡

† Other thermodynamic potentials could be considered such as the Helmholtz potential, $A(\varepsilon, M)$.

‡ A single variable function $f(r)$ is homogeneous if, for all values of the parameter λ, it satisfies

$$f(\lambda r) = g(\lambda) f(r) = \lambda^p f(r) \tag{82}$$

with p the so called degree of homogeneity. It follows, since it is always possible to put $r = \lambda r_0$, that

$$f(r) = \lambda^p f(r_0), \tag{83}$$

i.e. the function at any point r is related to its value at one point $f(r_0)$ by a simple change of scale (usually non-linear).

The definition can be extended to n variables. In particular for a homogeneous function of two variables $f(x, y)$, one has, for all values of the parameter λ,

$$f(\lambda x, \lambda y) = g(\lambda) f(x, y) = \lambda^p f(x, y) \tag{84}$$

A function $f(x, y)$ which satisfies the condition

$$f(\lambda^a x, \lambda^b y) = \lambda f(x, y) \tag{85}$$

is called a "generalized homogeneous function." Two undetermined parameters appear in it. Equivalent forms of (85) are

$$f(\lambda x, \lambda^b y) = \lambda^p f(x, y) \tag{86}$$

$$f(\lambda^a x, \lambda y) = \lambda^p f(x, y) \tag{87}$$

All these homogeneous functions have a particular feature, which we demonstrate for the case of a two variable function, starting from expression (85). Since (85) is valid for any

It is then possible to write

$$G(\lambda^{a_\varepsilon}\varepsilon, \lambda^{a_H}H) = \lambda G(\varepsilon, H) \tag{90}$$

i.e. for any value of the number λ, two parameters (a_ε, a_H) exist which make (90) valid.

It is possible to show that all critical exponents may then be expressed by means of these two parameters and to obtain definite predictions concerning the form of the state equation $M = M(\varepsilon, H)$. The fact *that the parameters* a_ε, a_H *are not specified, does not permit the numerical calculation of the exponents* but makes it clear that *only two exponents are really independent.*

To show the way in which exponents can be found we differentiate the two members of (90) with respect to H; remembering that $\left(\frac{\partial G}{\partial H}\right)_\varepsilon = -M$ one finds

$$\lambda^{a_H}M(\lambda^{a_\varepsilon}\varepsilon, \lambda^{a_H}H) = \lambda M(\varepsilon, H). \tag{91}$$

If we fix $H = 0$

$$M(\varepsilon, 0) = \lambda^{(a_H-1)}M(\lambda^{a_\varepsilon}\varepsilon, 0). \tag{92}$$

Since (92) is valid for all values of λ, let us choose $\lambda = \left(\frac{1}{\varepsilon}\right)^{1/a_\varepsilon}$. Then

$$M(\varepsilon, 0) = \varepsilon^{(1-a_H)/a_\varepsilon}M(1, 0). \tag{93}$$

Confronting (93) with the power law valid near the critical point $(\varepsilon \to 0^-)$

$$M(\varepsilon, 0) \sim \varepsilon^\beta \tag{94}$$

---

choice of λ, we put $\lambda = \frac{1}{y}$

$$f\left(\frac{x}{y}, 1\right) = y^{-p}f(x, y) \tag{88}$$

The function on the left hand side evidently becomes a function of the single variable $z = \frac{x}{y}$ and we obtain

$$f(x, y) = y^p F\left(\frac{x}{y}\right) \tag{89}$$

i.e. a homogeneous function depending upon x only through $\frac{x}{y}$. The converse is also true, i.e. a function that satisfies (89) is homogeneous. It is also possible to show that only functions of the type (89) (or of the same form with x and y interchanged) are homogeneous.

one obtains β as a function of the two parameters:

$$\beta = \frac{1 - a_H}{a_\varepsilon}. \tag{95}$$

In a similar way, returning to (91), fixing $\varepsilon = 0$ and letting $H \to 0$ one finds

$$\delta = \frac{a_H}{1 - a_H}. \tag{96}$$

From $\left(\frac{\partial^2 G}{\partial H^2}\right)$ (which is equal to $-\chi(\varepsilon, H)$) with $H = 0$ and $\varepsilon \to 0^-$ the exponent γ' is obtained

$$\gamma' = \frac{2a_H - 1}{a_\varepsilon}. \tag{97}$$

The relations (95), (96), (97) can be combined to furnish

$$\gamma' = \beta(\delta - 1)$$

which is Widom's result already quoted (equation 80). This is the first example of a general type of result based on the scaling hypothesis, namely validating as an equality a relation which had before been established rigorously only as an inequality (equation 69).

If instead we keep $H = 0$ but let $\varepsilon \to 0^+$ we find for γ the same expression as for γ', i.e.

$$\gamma = \gamma' \tag{98}$$

The symmetry of the critical exponents about T_c, of which (98) is a particular case, *is a general result of the scaling hypothesis*. From $\left(\frac{\partial^2 G}{\partial \varepsilon^2}\right)_H$ (which is connected with C_H) one has the possibility of calculating α'

$$\alpha' = 2 - \frac{1}{a_\varepsilon}. \tag{99}$$

Combining this with the preceding expression in which a_ε enters we obtain

$$\alpha' + \beta(\delta + 1) = 2 \tag{100}$$

$$\alpha' + 2\beta + \gamma' = 2 \tag{101}$$

which validates as equalities the Griffith (67) and Rushbrooke (59) relations. Other exponents and relations among them can be deduced along the lines just sketched.

The homogeneity assumption on the form of the Gibbs potential has of course immediate consequences for the form of the state equation; as a matter of fact equation (91), being a relation among H, M, and T is the state equation. It is possible to show that, with the help of experimental results such a relation can have the form of a series expansion first proposed by Domb and Hunter (equation 81). The sum of this series can be put in a form (Griffith, 1967)

$$H = M^{\delta} h(\varepsilon M^{-(1/\beta)}) = M^{\delta} h(x) \tag{102}$$

with $x = \varepsilon M^{-(1/\beta)}$. $h(x)$ is defined for $-x_0 < x < \infty$, $x = -x_0$ being the equation of the phase boundary and $x = \infty$ corresponds to the critical isochore $M = 0$. The critical isotherm ($\varepsilon = 0$) corresponds to $x = 0$; the critical point, corresponding at the same time to $x = 0$ and $x = \infty$, is excluded from consideration. $h(x)$ is an analytic function in the range of definition.

If now we make a change of variables choosing as new variables $\dfrac{H}{M^{\delta}}$ and $\varepsilon M^{-(1/\beta)}$, expression (102) takes a form which should be valid for any system at all temperatures. It lends itself well to comparison with experiment because all experimental points for a system should fall on the same line, independently of the particular experimental conditions. Figure 28 (Vicentini-Missoni, 1970) gives the plot of $\log h(x)$ versus $\log \dfrac{x - x_0}{x_0}$ for Ni (data of Kouvel and Comly, 1968); Fig. 29 is the analogous curve (Vicentini-Missoni, 1971) for $CrBr_3$ (data of Ho and Lister, 1969) and Fig. 30 (Vicentini-Missoni) gives the analogous results for fluid He^4 (data of Roach, 1968). The comparison is in all cases very satisfactory.

1. *Kadanoff's approach to scaling*

Kadanoff has presented an argument, far from rigorous, which however gives some intuitive physical support to the scaling idea and it is therefore worth mentioning notwithstanding its limitations. A general result coming from the analysis of critical phenomena (for instance the Ising model theories) is that the various correlation functions which may be considered (density-density or spin-spin; energy density-energy density; energy-density-spin) in the critical region go exponentially to zero as a function of the distance with the same characteristic range ξ.

Let us therefore start our considerations by assuming that static critical phenomena are characterized by a single diverging correlation length $\xi(\varepsilon)$.

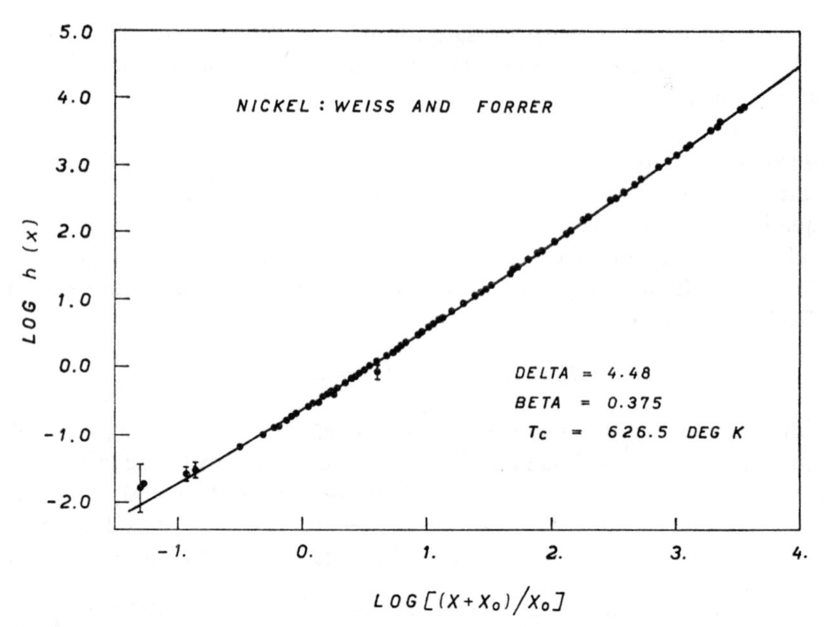

FIGURE 28. Scaling relation for Nickel (Vicentini and Missoni, 1970); experimental data are those of Kouvel and Comly (1968).

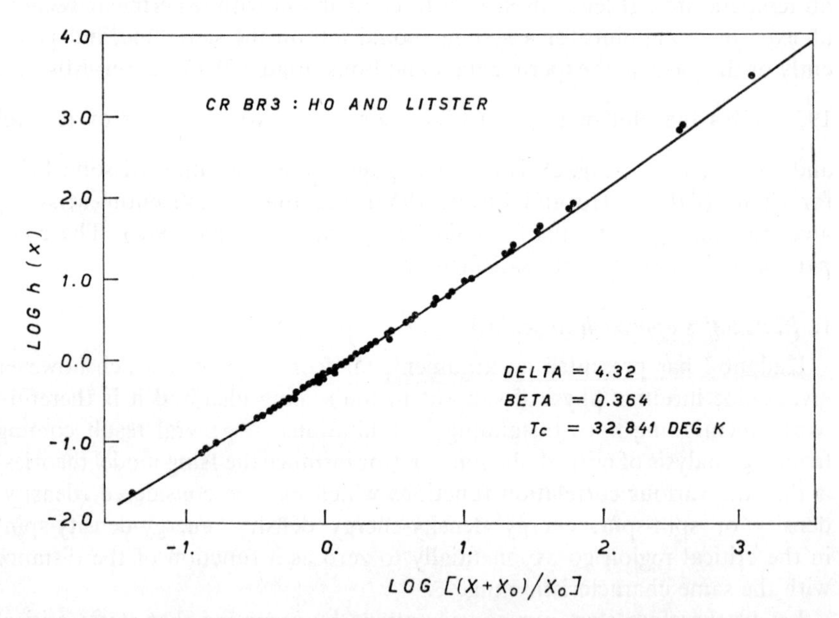

FIGURE 29. Scaling relation for the ferromagnet $CrBr_3$ (Vicentini and Missoni, 1971); the experimental data are those of Ho and Leister (1969).

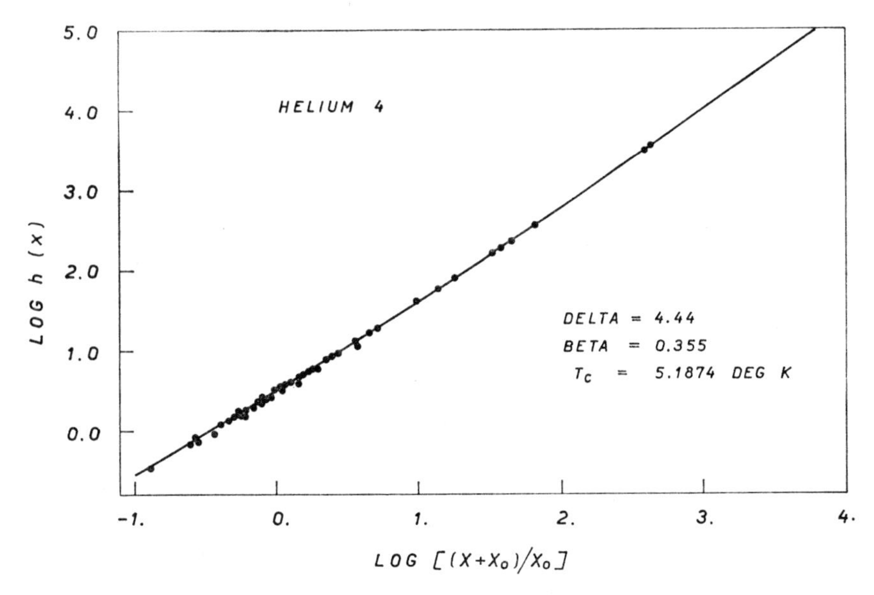

FIGURE 30. Scaling relation for He^4 (Vicentini and Missoni, 1971); the experimental data are Roach (1968).

This assumption seems plausible if one assumes that the interactions among particles which have importance in the transition of the particular system under consideration (which are mainly forces between neighbours) are essentially of one simple type, i.e. they are not forces of widely different types and ranges.

To introduce Kadanoff's scaling idea (1966) let us refer to an Ising model of a ferromagnet. The spins are at the sites of a regular lattice of constant a_0 (Fig. 31) and can take the two positions "up" or "down." The correlation length is very large in the critical region and Kadanoff's argument is valid only when $\xi \gg a_0$. Let us divide the crystal into cells of linear dimension La_0 (Fig. 31), L being arbitrary but such that

$$1 \ll L \ll \frac{\xi}{a_0} \tag{103}$$

i.e. in a correlated region (of dimensions of the order ξ^3) many, many cells fall. It is clear then that the interaction between adjacent cells must be of the utmost importance in producing the large range of correlation and in determining the critical behaviour of the physical system. Actually it should be almost equivalent to describing the critical behaviour of the physical system following either one of two descriptions of the system itself: the

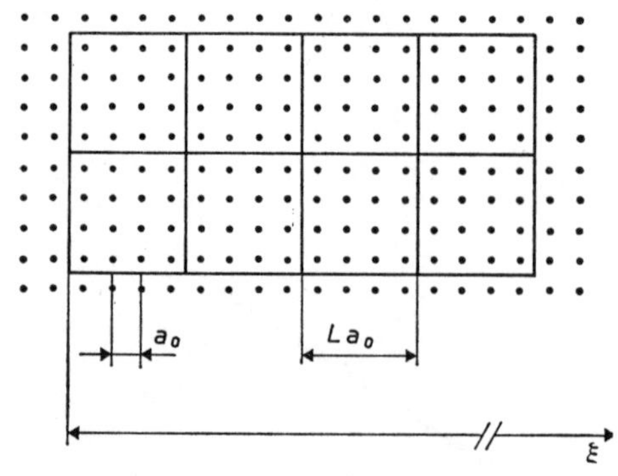

FIGURE 31. Quantities for the scaling law approach in an Ising model.

traditional one based on lattice sites and individual spins and one based on cells, assigning to each cell a magnetic moment which points "up" or "down." In other words, the very extended correlation among particles near the critical point is a process which does not depend strongly on the details of interactions between neighbours.

The Ising lattice problem is described by using a variable σ_r the spin variable, which depends upon the lattice site (**r**), a reduced magnetic field h, and the temperature (ε). The cell problem is described by a magnetic moment variable, μ_a which depends on the cell position (**a**) and on parameters $\tilde{\varepsilon}$, $\tilde{h}$ of the same type as those of the lattice problem. From what has been said about correlations, the two descriptions being equivalent at the critical point, $\tilde{\varepsilon}$ and $\tilde{h}$ should be proportional to ε and h and, following Kadanoff, we may assume the dependence on the dimension of the cell has the following form

$$\tilde{h} = L^x h$$

$$\tilde{\epsilon} = L^y \varepsilon \tag{104}$$

with x and y two new parameters.

Kadanoff then proceeds to calculate the variation of the Gibbs free energy for an infinitesimal variation of the magnetic field by using either the lattice or the cell description. This allows him to link μ_a to σ_r:

$$\sigma_r = L^{x-d} \mu_a \tag{105}$$

d being the dimensionality of the system. Moreover, for the case in which the field h_r does not depend on r, the average $\langle\sigma\rangle$ and $\langle\mu\rangle$ referring both to the identical problem must have the same functional expression, respectively

$F(\varepsilon, h)$ and $F(\tilde{\varepsilon}, \tilde{h})$. One can then write

$$\langle \sigma \rangle = F(\varepsilon, h) = L^{(x-d)} F(L^y \varepsilon, L^x h). \tag{106}$$

It has to be noted now that L is an irrelevant parameter which has been introduced only for calculation and should disappear on the right-hand side expression. This is possible only if $F(\varepsilon, h)$ has the form:

$$\langle \sigma \rangle = F(\varepsilon, h) = \frac{h}{|h|} |\varepsilon|^{(d-x)/y} f\left(\frac{\varepsilon}{|h|^{y/x}}\right) \tag{107}$$

where $f(z)$ is an unknown function. This however does not prevent us from determining the critical index as a function of x and y by differentiation or integration of $\langle \sigma \rangle$ with respect to ε or h.

All this treatment is equivalent to saying that the Gibbs potential in the lattice description [($G(\varepsilon, h)$, per site] and the Gibbs potential in the cell description [$G(\tilde{\varepsilon}, \tilde{h})$, per cell] should be related in the following way

$$G(\tilde{\varepsilon}, \tilde{h}) = L^d G(\varepsilon, h) \tag{108}$$

i.e.

$$G(L^y \varepsilon, L^x h) = L^d G(\varepsilon, h). \tag{109}$$

We recall that L must satisfy the condition $1 \ll L \ll \frac{\xi}{a_0}$. If however one assumes that (109) is valid for all values of L, it expresses the condition for $G(\varepsilon, h)$ to be a generalized homogeneous function. The similarity between Kadanoff's treatment of scaling and the homogeneous function approach is evident.

We have seen previously the agreement between the scaling law and some experimental results on state equations. To make a further comparison with the indications of model calculations and the experimental results on exponents, we follow Kadanoff *et al.* (1967). The scaling calculations give for the indices

$$\alpha = \alpha' = 2 - \frac{d}{y} \tag{110}$$

$$\beta = \frac{d - x}{y} \tag{111}$$

$$\gamma = \gamma' = \frac{d}{y} - 2\beta \tag{112}$$

$$\beta(\delta + 1) = \frac{d}{y}. \tag{113}$$

From the scaling assumption for correlation functions (see next paragraph)

$$\nu = \nu' = \frac{2 - \alpha}{d} \tag{114}$$

$$\frac{d\gamma}{2 - \eta} = 2 - \alpha . \tag{115}$$

It follows therefore that the combinations of parameters

$$(2 - \alpha), (2 - \alpha'), d\nu, d\nu', \frac{d\gamma}{2 - \eta}, (\gamma + 2\beta), (\gamma' + 2\beta), \beta(\delta + 1)$$

all represent the same ratio $\frac{d}{y}$. A comparison between this theoretical result and the results of calculations on models and of experiments has been made by Kadanoff *et al.* (1967) and is given in Table 4.

The values of y and $\Delta = \frac{2x}{y}$ are calculated from the exponents. The agreement for the 2-*d* Ising model case is complete. In the case of the 3-*d* Ising model the value of the combination of exponents is only approximately constant. The uncertainties in the exponents, due to differences in the various calculations, are not sufficient to explain the deviation from a single value. Passing to comparison with experiment, one finds that the values for the gas-liquid case seem to agree with the scaling predictions and with the values of the 3-*d* Ising model, to the limit of experimental accuracy.

In the comparison with ferromagnetic materials some results have not been considered, e.g. those in Ni which give $\beta \simeq 0.5$ and those in $YFeO_3$ which give $\gamma' \simeq 0.7$, $\delta \simeq 2.8$; these results are in contrast with the value obtained in the majority of ferromagnets ($\beta = 0.33 \pm 0.03$; $\gamma = 1.33 \pm 0.03$; $\nu = 0.65 \pm 0.03$; $\delta = 4.1 \pm 0.1$). The value of exponents in the exceptional cases just quoted are at times very near to those given by the classical theory and could be explained by assuming that long-range forces, such as magnetic dipole interactions, act in these materials when a substantial spin alignment exists. In such cases the scaling idea should not be applicable.

The comparison of Table 4 seems to indicate that scaling furnishes indications which are often, though not always, in satisfactory agreement with experiment and with model calculations. Frequently the deviations, although significant, are small enough to permit one to suppose that more sophisticated applications of the idea of scaling may eliminate them. At times however the deviations are so large as to indicate limitations in the applicability of the idea. For instance, improved numerical calculations of series expansion for

Table 4
Comparison of values of different critical exponents combinations

	$2 - \alpha$	$2 - \alpha'$	$d\nu$	$d\nu'$	$d\gamma/(2 - \eta)$	$\gamma + 2\beta$
2-*d* Ising model	2	2	2	2	2	2
3-*d* Ising model	1.87 ± 0.12	$1.93 {+0.04 \atop -0.16}$	1.93 ± 0.01		1.933 ± 0.008	1.87 ± 0.01
ferromagnets	1.92 ± 0.08	1.92 ± 0.08	1.95 ± 0.09		2.08 ± 0.12	1.99 ± 0.09
antiferromagnets	1.92 ± 0.08	1.92 ± 0.08	1.95 ± 0.09			1.96 ± 0.10
gas-liquid	1.8 ± 0.2	1.88 ± 0.12				2.06 ± 0.2

	$\gamma' + 2\beta$	$\beta(\delta + 1)$	y	$\Delta = 2x/y$
2-*d* Ising model	2	2	1	3.75
3-*d* Ising model	1.94 ± 0.05	1.93 ± 0.05	1.55 ± 0.01	3.22 ± 0.02
ferromagnets		1.7 ± 0.02	1.54 ± 0.07	3.2 ± 0.2
antiferromagnets			1.54 ± 0.07	3.2 ± 0.2
gas-liquid	1.7 ± 0.3	1.87 ± 0.14	1.57 ± 0.03	3.12 ± 0.1

the Ising model seem to give definite evidence for violation of the scaling result

$$d\nu + \alpha = 2 \tag{114}$$

and introduce some limitations on the validity of the general expression of the order parameter correlation function (58) [Moore *et al.*, 1969; Ferer *et al.*, 1969; Moore, 1970]. An attempt to understand such a limitation in the framework of a microscopic approach has been given by Migdal (1971).

These conclusions together with those reached in the preceding section explain why scaling, although based on unproved hypotheses has created a great deal of interest and is often used to obtain results for use in other calculations.

H. SCALING FOR THE PAIR CORRELATION FUNCTION

Kadanoff has applied his scaling approach to the spin-spin correlation function. In the lattice scheme this function† is

$$G(\mathbf{r}, \mathbf{r}') = G(r, \varepsilon, h) = \langle[\sigma_r - \langle\sigma\rangle][\sigma_{r'} - \langle\sigma\rangle]\rangle, \tag{116}$$

$$r = \frac{|\mathbf{r} - \mathbf{r}'|}{a_0}$$

being the distance between the two points which are considered. In the cell scheme the correlation function must have the same expression except for variation in the scale of length, ε and h, namely r is to be changed to $\frac{r}{L}$; $\varepsilon \rightarrow \tilde{\varepsilon} = \varepsilon L^y$; $h \rightarrow \tilde{h} = hL^x$. We may write, by using (105)

$$\begin{aligned} G(r, \varepsilon, h) &= L^{2(x-d)}\langle[\mu_a - \langle\mu_a\rangle][\mu_a' - \langle\mu_a\rangle]\rangle \\ &= L^{2(x-d)}G\left(\frac{r}{L}, \varepsilon L^y, hL^x\right). \end{aligned} \tag{117}$$

Again, L has been introduced only for calculation and the function G should be independent of L. This specifies the form of the second member

$$G(r, \varepsilon, h) = |\varepsilon|^{2(d-x)/y}\, G\left(r, |\varepsilon|^{1/y}, \frac{\varepsilon}{|h|^{y/x}}\right) \tag{118}$$

† In this paragraph G is the symbol for the correlation function and not for the Gibbs potential.

for $r \gg 1$, $|\varepsilon| \ll 1$, and $h \ll 1$. From expression (118) one finds immediately that for $h = 0$ the coherence length is proportional to $|\varepsilon|^{-1/y}$ and therefore

$$\nu = \nu' = \frac{1}{y} = \frac{2-\alpha}{d}. \tag{119}$$

Similarly, because for $\varepsilon = 0$ and $h = 0$, and large r (equation 58)

$$G(r) \sim (r^{d-2+\eta})^{-1}$$

one gets

$$\frac{d\gamma}{2-\eta} = 2 - \alpha. \tag{120}$$

Kadanoff's approach is again very near to the assumption of homogeneity for the correlation function. In fact let us write equation (117) for $h = 0$

$$G(r, \varepsilon) = L^{2(x-d)} G(rL^{-1}, \varepsilon L^{y}). \tag{121}$$

If we assume that (121) is valid for all values of L, this is the same as assuming that $G(r, \varepsilon)$ is a generalized homogeneous function.

The scaling argument applied to the correlation function allows us to express exponents ν and η, through x and y, in terms of the exponents ($\alpha, \beta, \gamma, \ldots$) introduced in the description of the behaviour of the thermodynamic derivatives. This opens the way to establishing as equalities relations in which exponents of the two types enter and which were already written rigorously as inequalities.

Another approach to the scaling of the correlation functions has been given by Halperin and Hohenberg (1969); we will return to it in the treatment of dynamic phenomena.

I. UNIVERSALITY HYPOTHESIS AND SCALING

At various times it has been pointed out how frequently one observes similarities of critical processes occurring both in systems of the same type and in systems of different types, in addition to the equality, within the limits of experimental accuracy, of critical exponents for physical quantities which play an analogous role in the different transitions. The idea of the existence of basic common properties bound to the statistical nature of critical processes and to the role of fluctuations emerges naturally and leads to the *universality hypothesis*. In trying to illustrate universality we will follow Kadanoff (1970) who has also linked this hypothesis to scaling: the considerations which will be given also complete nicely the observations made at various points earlier.

The description of a phase transition requires the use of a free energy as a function of thermodynamic variables. The system is usually defined by means of two pairs of conjugate variables, one intensive (or field) and one extensive. Normally one field variable, such as the magnetic field, h, in ferromagnets or the pressure p or the chemical potential, μ, in simple fluids, is the variable which drives the system from one coexisting phase to the other and has as conjugate variable the order parameter (M, $\rho - \rho_c$). The second field variable is usually $\varepsilon = \dfrac{T - T_c}{T}$ which is used to bring the system towards or away from the critical point. Table 5 gives the two pairs of thermodynamic variables commonly used for a few phase transitions. The free energy is a function of the two intensive variables, say h, ε

$$F(\varepsilon, h).$$

The two extensive variables appear in the differential dF

$$\mathrm{d}F = M\,\mathrm{d}h + \langle\mathfrak{H}\rangle\,\mathrm{d}\varepsilon. \tag{122}$$

In most cases the conjugate of ε, i.e. $\langle\mathfrak{H}\rangle$ has the interpretation of an energy.

Let us consider the second derivatives. In the magnetic case they express respectively: (1) $\dfrac{\partial^2 F}{\partial h^2}$ the magnetic susceptibility; this can be expressed by means of fluctuations of the order parameter $\langle(\partial M)^2\rangle$.

(2) $\dfrac{\partial^2 F}{\partial\varepsilon\,\partial h}$ the temperature dependence of the magnetization; this can be expressed by means of $\langle\delta M\,\delta\mathfrak{H}\rangle$.

(3) $\dfrac{\partial^2 F}{\partial\varepsilon^2}$ the specific heat at constant field, this can be expressed by means of fluctuations in the energy $\langle\delta\mathfrak{H}\,\delta\mathfrak{H}\rangle$.

Table 5
Conjugate variables used in various critical systems.

Phase transition	first pair order parameter	conjugate	second pair	
Ferromagnetic	M_z	H_z	$\mathfrak{H}$	ε
antiferromagnetic	sublattice magnetization	staggered magnetic field	$\mathfrak{H}$	ε
liquid-gas	$\rho - \rho_c$	$\mu - \mu_c$	$\mathfrak{H} - \mu\rho$	ε

In other phase transitions the derivatives give rise to analogous quantities (for simple fluids K_T, C_v). It is important to observe that in all cases the order parameter fluctuations are very strong and cause the second derivative with respect to the conjugate variable to be highly singular at the critical point. The weaker fluctuations are in energy ($\mathfrak{H}$) and this leads to a very weak singularity in the specific heat. These similarities among systems are reflected in the magnitudes and values of critical exponents for the corresponding quantities, as indicated in Table 6. These striking similarities among critical processes of different kinds suggest that the details of the system undergoing

Table 6
Critical exponents for the second field derivatives of the free energy in various systems.

Phase transition	Strong singularity	Intermediate singularity	Weak singularity
Mean field theory	1	1/2	0
$d = 2$ Ising	7/4	7/8	0
simple fluids	1.25	0.65	0.1
$d = 3$ Ising	1.25	0.685	0.125

transition may be unimportant. This leads to the formulation of the *hypothesis of universality* according to which phase transitions can be divided into a small number of classes according to the symmetries of the ordered state and the dimensionality of the system: all phase transitions in each class are identical and only the names of the variables are needed to describe the system change. The origin of this universality can be further exposed if one stops to consider a little more the links between some thermodynamic quantities and the fluctuations of the order parameter. If for instance we direct attention to the magnetic susceptibility, this is

$$\chi = \langle(\delta M)^2\rangle. \tag{123}$$

Because the magnetization can be expressed as an integral over all space of the magnetization density

$$M = \int \mathrm{d}r m(\mathbf{r}) \tag{124}$$

one has

$$\chi = \int \mathrm{d}r\, \mathrm{d}r' \langle \delta m(\mathbf{r})\, \delta m(\mathbf{r}')\rangle \tag{125}$$

and the susceptibility per unit volume

$$\frac{\chi}{V} = \int \mathrm{d}r \langle \delta m(r)\, \delta m(0) \rangle \,. \tag{126}$$

In equation (126) one has to observe that the magnetization density remains finite and so the same is true of the integrand. If therefore the susceptibility, i.e. the integral, is found to diverge at the critical point, it is clear that the only way in which this may occur is that the range of the integrand becomes larger and larger as the critical point is approached.

The conclusion is therefore reached that critical phenomena (such as the divergence of χ or similar quantities) are due to the fact that the range (or correlation length, ξ) of the order parameter fluctuations becomes infinite as the critical point is approached. We have at various times indicated how experiments and their interpretations point to this conclusion. The physical reason for this growth of fluctuations lies in an enormous increase, both in number and in spatial size, of heterophase embryos, always present in a phase, when the free energy required for their creation becomes gradually smaller and smaller until it disappears.

The basis of the universality hypothesis is in the fact that critical phenomena depend on the fluctuations of quantities such as M and $\mathfrak{H}$ and these fluctuations, as the critical point is approached, increase so much in range that they become insensible to the details of the interatomic potential; they only see some coarse characteristics of the potential such as the amount of breaking of exact symmetry (measured by h_z) or the distance to the critical point (measured by ε). The nature of the fluctuations is instead strongly dependent on the symmetries of the order parameter and the dimensionality of the system.

A mathematical formulation of the phase transition problem in which the universality hypothesis can be easily expressed is the following, proposed by Kadanoff. Let us introduce in the free energy function another field (λ) and its thermodynamic conjugate U which describe an internal characteristic of the particular system

$$F = F(\varepsilon, h, \lambda) \tag{127}$$

Continuous variation of λ from 0 to 1 can for instance be associated in the Hamiltonian of the system with passage from the Ising model to the Heisenberg model, or from a nearest neighbour interaction to a next nearest interaction. The variation of λ leaves unchanged the basic symmetry of the system and corresponds to the passage from one system to another which undergoes a particular class of phase transitions.

The universality hypothesis then is equivalent to assuming that λ is irrelevant in determining the basic thermodynamic functions and the correlation functions, i.e. that these depend on λ only via a trivial change of variables.

In such a case the functional forms at $\lambda \neq 0$ are the same as at $\lambda = 0$ except for a multiplication of the variables (such as h, ε, M) by constants (a, b, c) depending on λ.

The form of the correlation functions and thermodynamic functions as well as the critical indices remain the same for various λ, i.e. for the different phase transitions in a class. A close connection between the universality hypothesis and scaling can be demonstrated. Kadanoff has shown that scaling in his formulation (see Section II.H) can be deduced from universality as the case in which the *irrelevant* parameter λ is the scale of length in the problem.

Limitations to the universality hypothesis can be conjectured. For example Baxter (1972) in a particular model has found evidence for critical indices depending on the parameter λ.

At the end of the presentation made in these paragraphs of the general ideas which, although yet in formation, are little by little producing a physical understanding of static critical phenomena, we wish to point out how these ideas have been introduced on the basis of phenomenological considerations and have been checked by means of model calculations; they have not yet received a microscopic justification. An attempt to construct a theory of static critical behaviour on the basis of the general methods of statistical mechanics is now growing. The universality hypothesis has been reformulated on microscopic grounds by Kadanoff (1969a, b). General field theory methods have been quite successfully introduced (Polyakov, 1969, 1970; Migdal 1969). A review of these attempts can be found in de Pasquale *et al.* (1970).

This field theory approach to phase transitions has recently yielded a new method for calculating critical indices and quite good estimates of them (Wilson and Fisher, 1971; Wilson, 1971; de Pasquale and Tombesi, 1972).

III. Dynamic Critical Processes

A. INTRODUCTION

Non-equilibrium behaviour of critical systems may appear in a large variety of experiments which range from normal macroscopic transport processes occurring when the system is exposed to a macroscopic gradient of a thermodynamic variable, to phenomena such as sound absorption, relaxation processes, inelastic scattering of light or neutrons where either fluctuations of the thermodynamic variables are introduced in a system macroscopically in equilibrium or the magnitude and time decay of thermal fluctuations are studied. The last experiments allow the investigation of the transport properties as a function of the wave number k and the frequency ω.

The dynamic critical behaviour of a system is much more complex than the equilibrium one. An important point to be noted is that, in contrast to what has been found for equilibrium properties, significant differences appear to exist much more frequently in the dynamic behaviour of critical systems of different kinds.

This indicates that the particular nature of the interactions among particles have here a much more direct influence than for the equilibrium properties which, as we have already seen, are essentially determined by the statistical nature of the processes.

We will limit our consideration to fluid systems and will consider in the next sections typical results furnished by "thermodynamic" experiments, i.e. experiments performed with macroscopic gradients, by sound propagation investigations and will discuss at some length the kind of information obtainable by scattering experiments.

A short account of present theoretical ideas and procedures for tackling the problem of describing the dynamic behaviour of critical systems follows.

B. THERMODYNAMIC TRANSPORT EXPERIMENTS. SHEAR VISCOSITY

The transport coefficients are introduced as *constants* in the relations between irreversible flows and thermodynamic forces. The coefficient of thermal conductivity, λ,† enters in Fourier's heat conduction law

$$\mathbf{q} = -\lambda \nabla T$$

relating the heat flux $\mathbf{q}$ and the temperature gradient. The diffusion constant D appears in the diffusion law relating the flux of chemical species i, $\mathbf{J}_i$, and the concentration c_i

$$\mathbf{J}_i = -D \nabla c_i \tag{128}$$

The shear viscosity coefficient, η,‡ and the second viscosity coefficient η', enter in the Newton-Cauchy-Poisson law for the stress tensor (τ_{ij}) in an isotropic viscous fluid as a (linear) function of the rate of deformation:

$$\tau_{ij} = -p\,\delta_{ij} + \eta'\,\mathrm{d}_{kk}\,\delta_{ij} + 2\eta\,\mathrm{d}_{ij}; \qquad (\tau_{ij} = \tau_{ji}) \tag{129}$$

where p is the hydrostatic pressure (the minus sign appears because a tension is considered as a negative pressure in a fluid); δ_{ij} is the Kronecker symbol which is equal to 1 when $i = j$ and zero otherwise; d_{ij} is the rate of deformation tensor

$$d_{ij} = \frac{1}{2}\left(\frac{\partial u_i}{\partial x_j} + \frac{\partial u_j}{\partial x_i}\right) \tag{130}$$

† The symbol λ is used; no confusion should arise with other meanings of the symbol in the text (eq. (127), wavelength, λ point).

‡ The symbol η is used; no confusion should arise with the critical exponent of the net pair correlation function.

and **u** the velocity of the material particles. The term bulk viscosity coefficient η_B, is usually used for the linear combination

$$\eta_B = \eta' + \tfrac{2}{3}\eta. \tag{131}$$

The shear viscosity coefficient can be determined in a "thermodynamic experiment" where the flow is observed by applying a shear stress

$$\tau_{xy} = -p_{xy} = \eta\left(\frac{\partial u_x}{\partial y} + \frac{\partial u_y}{\partial x}\right) \tag{132}$$

The second viscosity and the bulk viscosity coefficients can only be determined in a different kind of experiment such as ultrasonic or light scattering experiments.

A critical analysis of the results so far gathered for λ and η in fluids has recently been given by Sengers (1970). As an example of the behaviour of shear viscosity in a two component critical system as a function of concentration we give in Fig. 32 the results of Leister *et al.* (1969) in mixtures of 3-methylpentane and nitroethane: the dotted lines give an estimated variation of η in the absence of the critical anomaly. Such an anomaly increases as the critical temperature is approached. The shape of such an anomaly as a function of concentration changes from system to system. It appears therefore more useful to study the temperature dependence of η at a fixed concentration. This is done by plotting log η as a function of T (or ε). Arrhenius equation

$$\eta = \eta_0 e^{E/RT} \tag{133}$$

should then lead to a linear relation in the proximity of the critical temperature ($\varepsilon \ll 1$). As Fig. 33 and 34 show, such a behaviour is indeed found for composition far from the critical one while an anomaly appears at the critical composition. The presence of such an anomaly in binary critical mixtures is at the present an experimentally well established result. The experiment has also confirmed the independence of the viscosity coefficient from the shear rate used. The separation of the critical anomaly ($\Delta\eta$) from the non-critical contribution (η_{id}) in the experimental results for η at the critical concentration

$$\eta = \Delta\eta + \eta_{id} \tag{134}$$

and the determination of the temperature dependence of $\Delta\eta$ has been the major effort of recent research. Sengers' (1970) analysis of these results at the critical concentration shows that by assuming for $\Delta\eta$ an expansion

$$\Delta\eta = H\varepsilon^{\phi} + \cdots \tag{135}$$

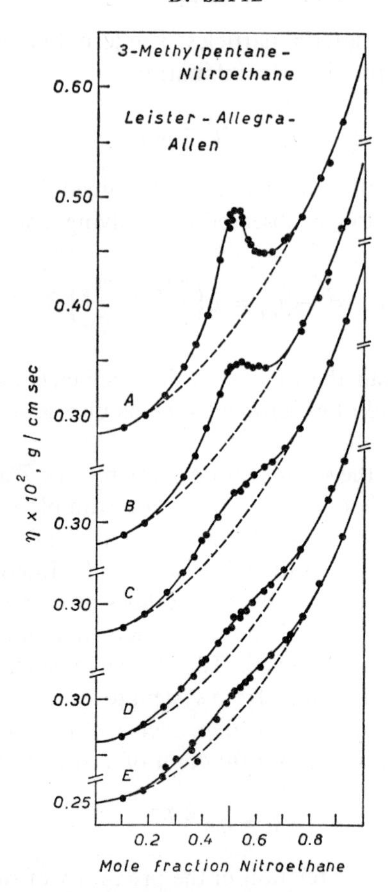

FIGURE 32. Shear viscosity in mixtures of 3-methylpentane as a function of concentration (Leister *et al.* (1969).)

of which only the first term need be considered, the critical exponent ϕ lies somewhere between -0.1 and $+0.1$. The anomaly could be a weak divergence, close to a logarithmic one or have a cusp-like behaviour. Such a result is definitely in contrast with the indication of some early (mean field) theories ($\phi = -\frac{1}{2}$) while it seems in agreement with recent theories (see Section I).

There are also some theoretical as well as experimental indications that the critical exponent should be determined not for $\Delta\eta$ but for the relative anomalous viscosity

$$\frac{\Delta\eta}{\eta_{id}} = H'\varepsilon^{\phi}. \tag{136}$$

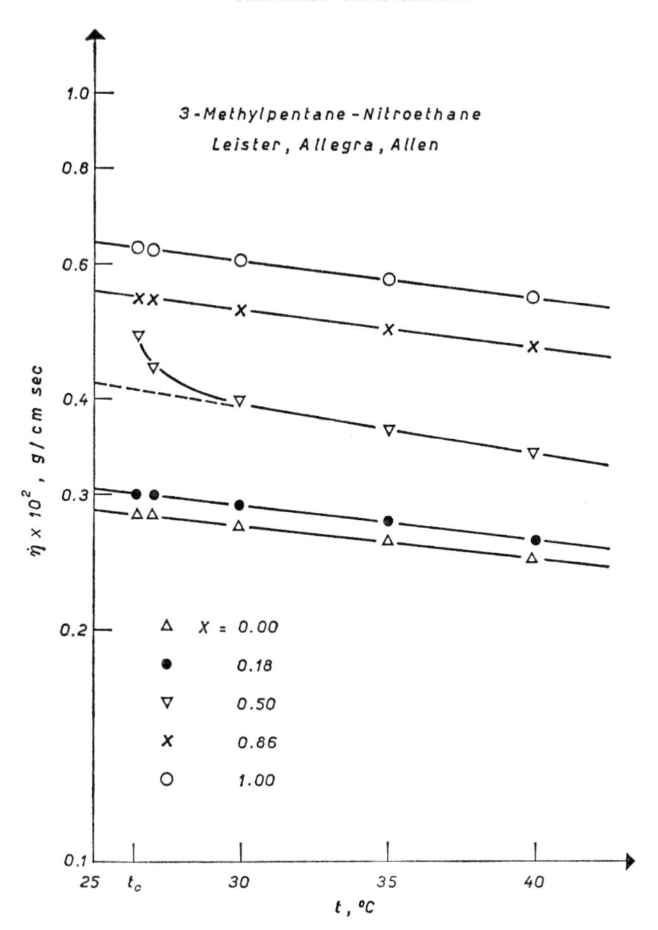

FIGURE 33. Log η vs temperature for different mole fraction (x), of nitroethane and 3-methylpentane (Sengers, 1970); the data are those of Leister *et al.* (1969).

Because η_{id}, determined for instance by means of the Arrhenius equation, depends on temperature, equation (136) is equivalent to using the series (135), with an estimate for the second order term in the expansion.

The measurements of shear viscosity near the gas-liquid critical point and their analysis are much more difficult: the accuracy of the measurements is usually no better than a few percent and the gravitational effect may introduce appreciable differences between the local (e.g. at the level of the oscillating disk if the viscosity is measured through the damping of an oscillating disk) and the average density. Figure 35 gives the viscosity results in CO_2 of Naldrett and Maass (1944) plotted as a function of local densities. These were estimated by Sengers by using a scaled equation of state for finding the density

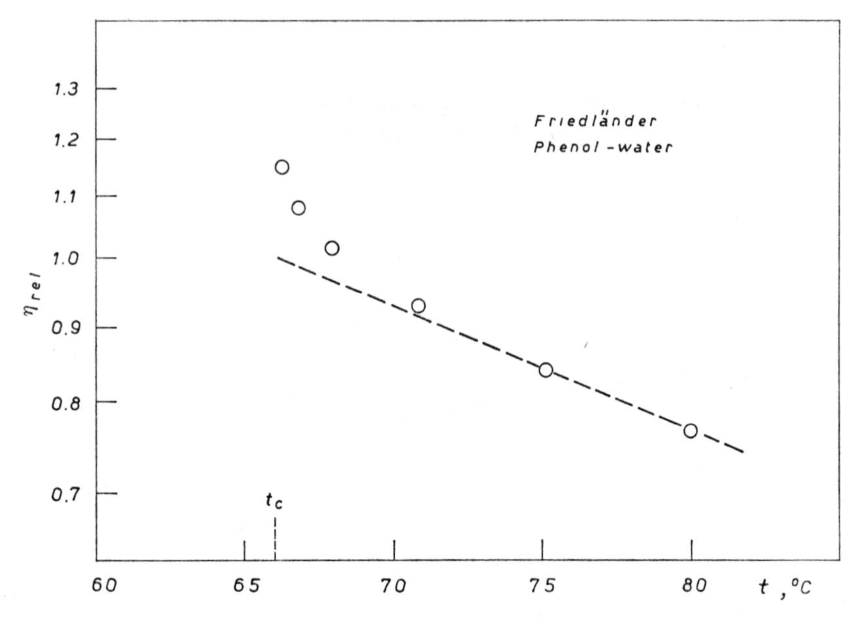

FIGURE 34. log η_{rel} vs temperature for phenol-water critical mixture (Sengers, 1970); the data of viscosity relative to the viscosity of water at 25°C are those of Friedlander (1901).

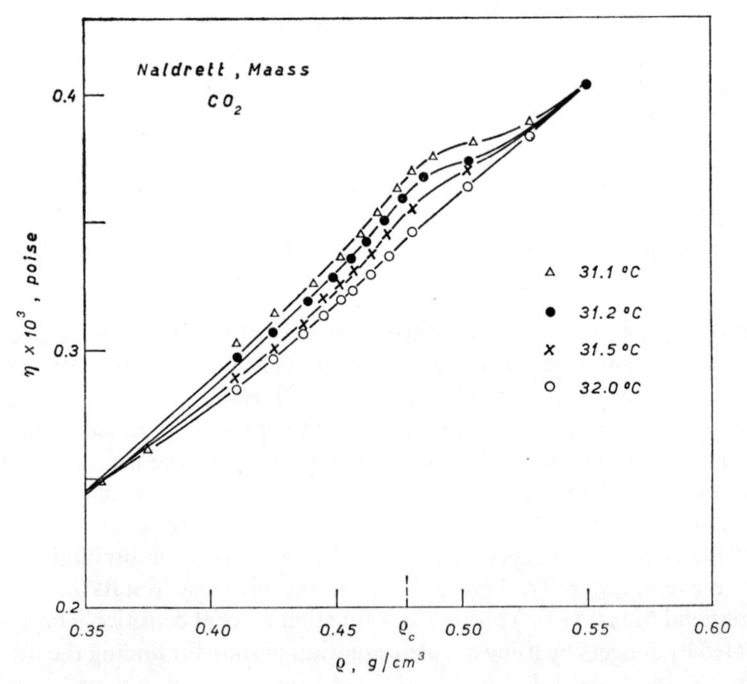

FIGURE 35. Viscosity of CO_2 as a function of the estimated local density (Sengers, 1970); viscosity data of Naldrett and Maass (1944).

profile in the vessel. Evidence of this kind indicates the existence of a weak anomaly when $\varepsilon < 10^{-2}$. The precisions of viscosity measurements and of the critical temperature determination do not permit for the present a specification of critical exponents for such an anomalous viscosity ($\Delta\eta = \eta - \eta_{id}$).

C. THERMAL CONDUCTIVITY AND DIFFUSION

Thermal conductivity measurements in the critical region of a gas-liquid system which are free from disturbance due to heat convection† are rather difficult. The analysis of the results is frequently made still more difficult by an inadequate specification of the density at which the thermal conductivity is measured. Results in convection free experiments in CO_2 by Michels and Sengers (1962) are given in Fig. 36. Experiments in argon, nitrogen, para-hydrogen, He^3, ammonia and in other single fluid critical systems show similar results. Figure 37 gives, for the CO_2 measurements of Fig. 36, the anomalous part of λ:

$$\Delta\lambda = \lambda - \lambda_{id}$$

as evaluated by Sengers (1970). The results on the critical isochore ($\rho = \rho_c$) can be represented by a power law

$$\Delta\lambda(\rho_c, T) = \Lambda\varepsilon^{-\psi} \tag{137}$$

with $\Lambda = (0.0030 \pm 0.0004)\dfrac{W}{m^\circ C}$, $\psi = -0.60 \pm 0.05$. Similar results on the thermal conductivity divergence are found for argon.

Thermal conductivity enters, as we will see shortly, in the Rayleigh line width of the scattered light which is proportional to

$$\frac{\lambda}{C_p}$$

The results obtained with this technique in general support a strong divergence of λ. While information on the thermal conductivity of critical binary liquid mixtures is not sufficient to draw any sure conclusions, exhaustive research has been carried out in liquid Helium I near the λ transition by Ahlers (1968). His detailed analysis of the results shows that the thermal

† The thermal expansion coefficient $\alpha_p = -\dfrac{1}{\rho}\left(\dfrac{\partial\rho}{\partial T}\right)_p$ diverges in the critical region like the isothermal compressibility. Finite temperature gradients which are necessary in the thermal conductivity measurements, therefore, lead to large density differences. These differences have not only the effect that λ changes through the cell, but they may introduce convection.

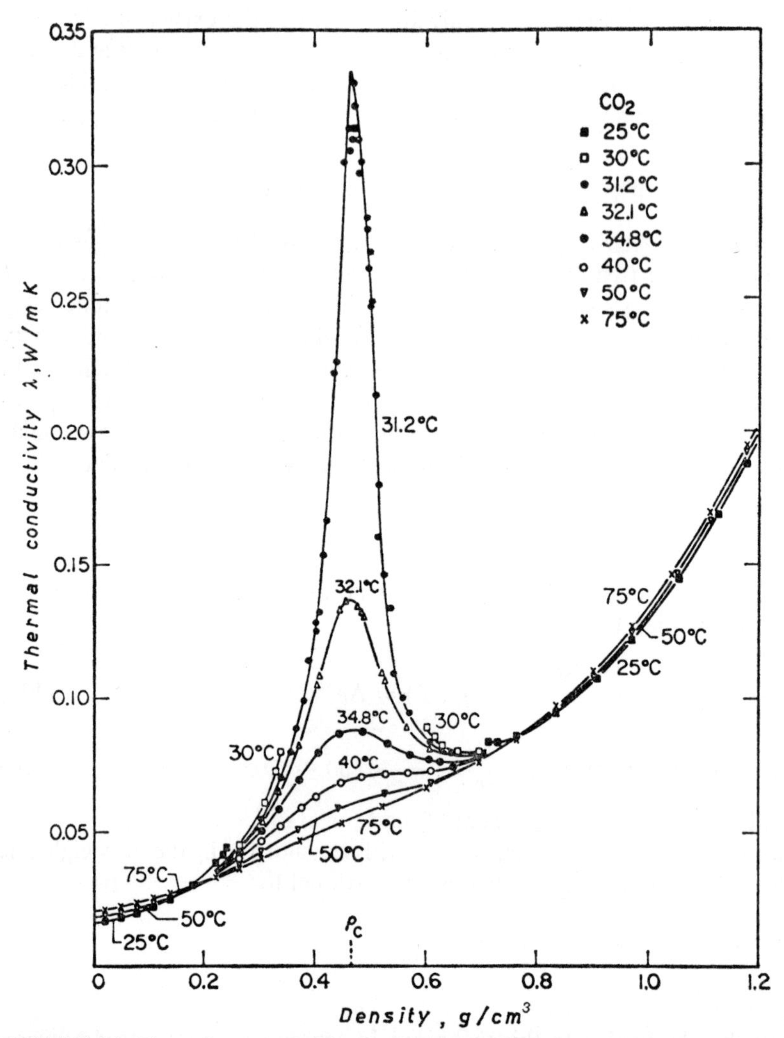

FIGURE 36. Thermal conductivity of CO_2 vs density at various temperatures (Michels and Sengers, 1962).

conductivity diverges as $\varepsilon^{-\psi}$ with $\psi = 0.334 \pm 0.005$: this conclusion agrees with the prediction of dynamic scaling theories ($\psi = \frac{1}{3}$).

Some information on diffusion constants has been obtained with various techniques ranging from N.M.R. to the study of mobility of ions in He, and from the observation by motion pictures of the dust particle motion in a fluid to light or neutron scattering. The experimental results are far from sufficient to draw clear conclusions. For instance some N.M.R. measurements

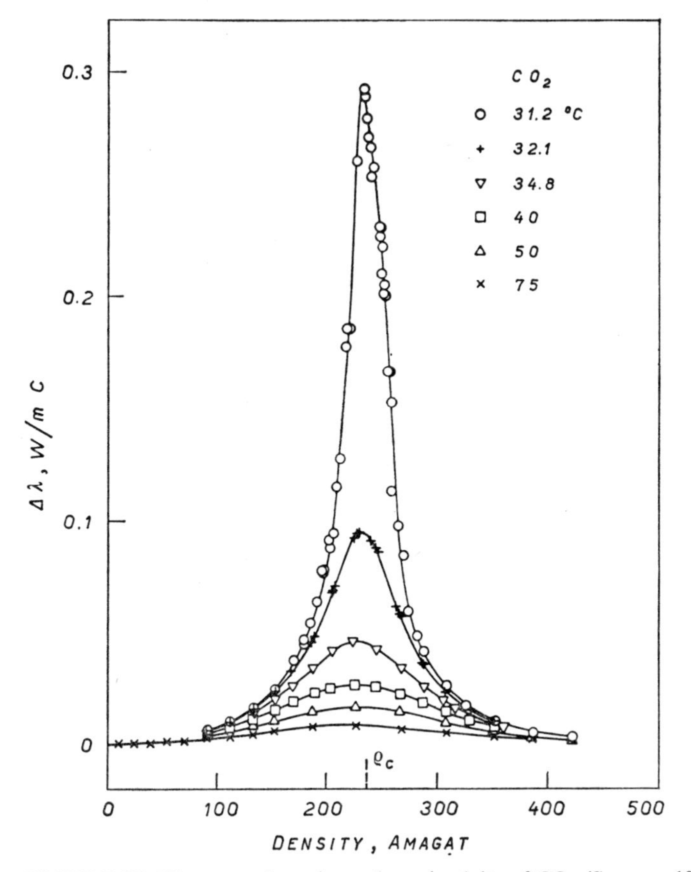

FIGURE 37. The anomalous thermal conductivity of CO_2 (Sengers, 1970).

in ethane (Noble and Bloom (1965)) seem to indicate a small anomaly at the critical point, others in methane (Trappeniers *et al.*, 1965, 1966), do not.

D. SOUND PROPAGATION

Sound propagation may be studied in a wide frequency range. In fluids, experiments can be performed from frequencies in the kilohertz region up to about 1 kilomegahertz with ultrasonic techniques. Measurements in the region between various hundreds of megahertz and a few kilomegahertz (hypersound) are also possible by observing the Brillouin scattering. This is the scattering of a parallel beam of light at an angle θ, produced by sound waves of thermal origin in which the thermal agitation in the liquid can be resolved according to Debye. We will discuss the Brillouin scattering in the next section.

The velocity of small amplitude sound waves in the zero frequency limit is connected with equilibrium thermodynamical properties of the system:

$$u^2 = \left(\frac{\partial p}{\partial \rho}\right)_s = \frac{1}{\rho K_s} \tag{138}$$

In a liquid-gas critical system one may write:

$$\rho u^2 \simeq T(\rho_c C_v)^{-1}\left(\frac{\partial p}{\partial T}\right)_{\text{sat}} \qquad T \to T_c\,, \tag{139}$$

$\left(\left(\frac{\partial p}{\partial T}\right)_{\text{sat}}\right.$ being the limiting value of the slope of the coexistence $\left.\text{curve}\right)$, and u^2 should go to zero as C_v^{-1} as $T \to T_c$. We will not discuss further here the information that zero frequency sound velocity has provided (Sette, 1970). We will instead consider sound absorption and velocity dispersion which are connected with the dynamic behaviour of the system. The absorption coefficient† of plane waves as calculated, in the Kirchhoff approximation, from hydrodynamics, gives

$$\alpha_{s,k} = \frac{\omega^2}{2\rho_0 u_0^{\,3}}\left[(\tfrac{4}{3}\eta + \eta_B) + (\gamma - 1)\frac{\lambda}{C_p}\right] \tag{140}$$

The contribution due to thermal conductivity is usually negligible. The determination of the low frequency limit of the experimental values of $\frac{\alpha}{f^2}$ (f being the frequency) provides a method for finding the hydrodynamic bulk viscosity. Both viscosity coefficients however, show a frequency dependence: in low (shear) viscosity systems the frequency variation of $\frac{\alpha}{f^2}$ can be ascribed to a frequency dependent bulk viscosity. This of course, is a global representation of multiple relaxation processes of various kinds occurring in the system where equilibria of different types are perturbed by the pressure and temperature variations associated with wave propagation. Velocity dispersion sets in for the same processes. The times involved in these relaxation phenomena determine the frequency regions in which $\frac{\alpha}{f^2}$ varies.

Figure 38 shows the existence of dispersion in the critical region of CO_2. The results obtained in the ultrasonic region (0.2–2 MHz) by various authors are compared (broken lines) with the measurements (solid lines) in the range

† The symbol α is used for the sound absorption coefficient, notwithstanding that it has been already used for the C_v critical exponent, no confusion being possible.

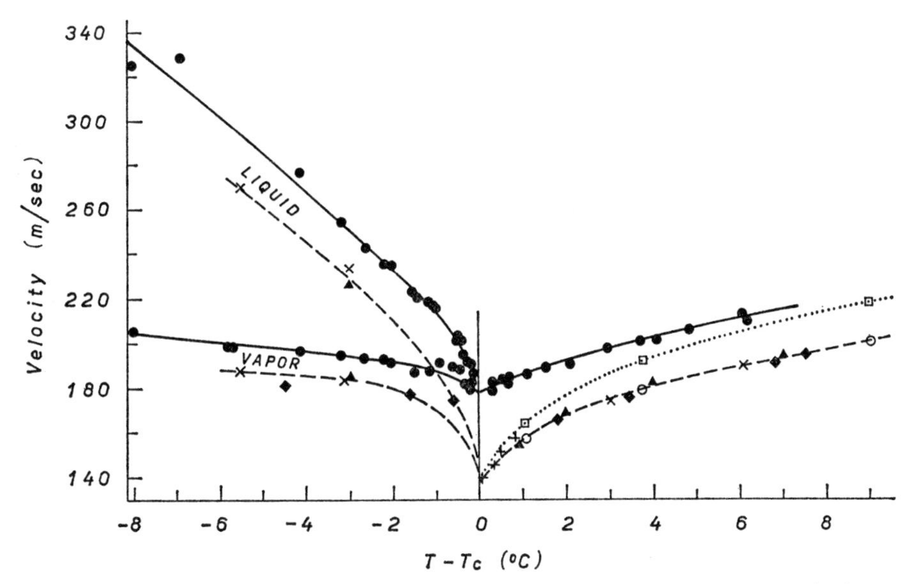

FIGURE 38. Sound velocity in CO_2: full lines give the hypersonic measurements; broken lines correspond to ultrasonic data of various authors; dotted line gives for $T > T_c$ the theoretical correction of low frequency velocity for vibrational dispersion (Gammon *et al.*, 1967).

425–840 MHz performed by Gammon *et al.* (1967) by means of the Brillouin scattering technique. In order to compare the ultrasonic data with the Brillouin scattering results and ascertain if there is a dispersion connected with critical effects, the dotted line has been calculated by taking into account the dispersion due to vibrational relaxation that has a relaxation frequency of about 10 MHz in gaseous CO_2 at densities of the order of ρ_c and corresponds to a phenomenon which is practically extinguished in the range 400–800 MHz. The difference between the dotted line and the solid line in the one phase region is the critical dispersion. Other determinations in xenon of critical dispersion will be considered later. Figure 39 gives the absorption coefficient per wavelength ($\alpha \cdot \lambda$) in xenon at 0.250 MHz as a function of the temperature (Chynoweth and Schneider (1952)). The absorption increases quickly when approaching the critical temperature in a temperature range of ± 2°C.

Figures 40 and 41 show how the parameter $\frac{\alpha}{f^2}$ changes with composition, frequency and temperature above the consolute temperature in the binary system formed by nitrobenzene and *n*-hexane (D'Arrigo and Sette, 1968). Figure 42 gives the variation of $\frac{\alpha}{f^2}$ with the temperature on both sides of the

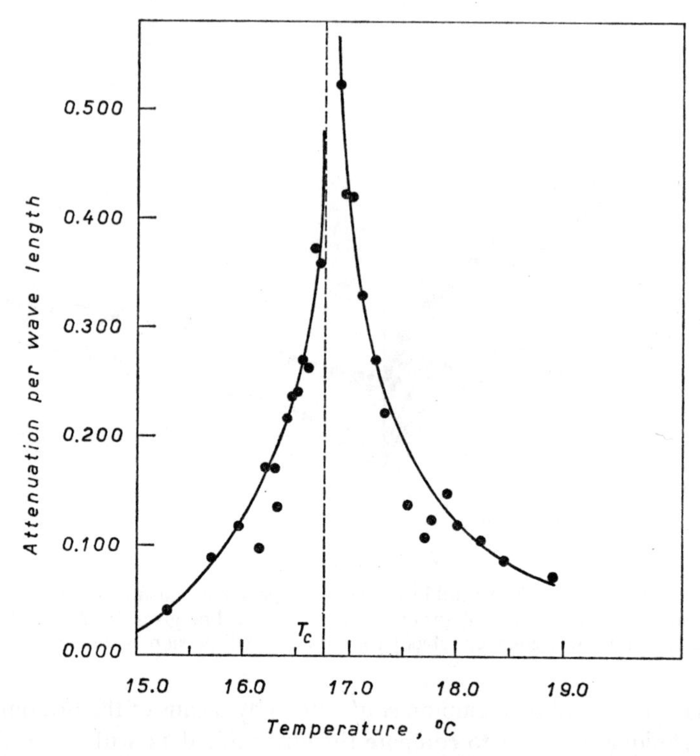

FIGURE 39. Sound absorption per wavelength in xenon at 0.250 MHz (Chynoweth and Schneider, 1952).

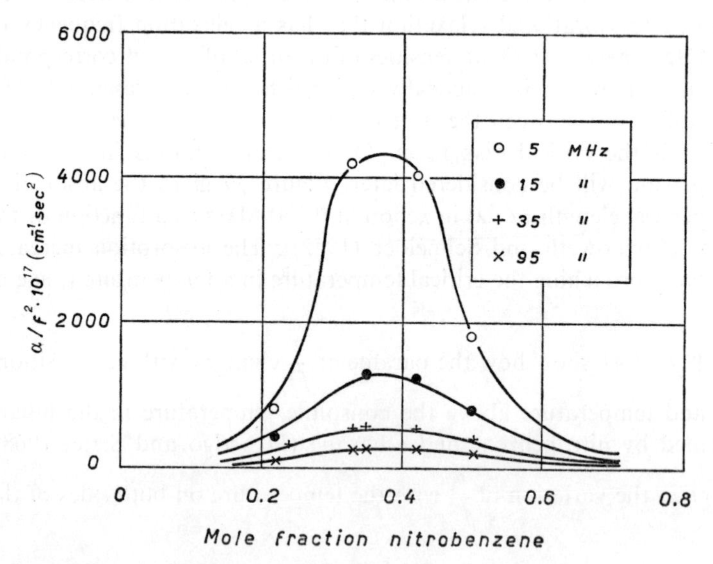

FIGURE 40. Sound absorption vs composition at same frequencies in nitrobenzene-*n*-hexane mixtures (22°C) (D'Arrigo and Sette, 1968).

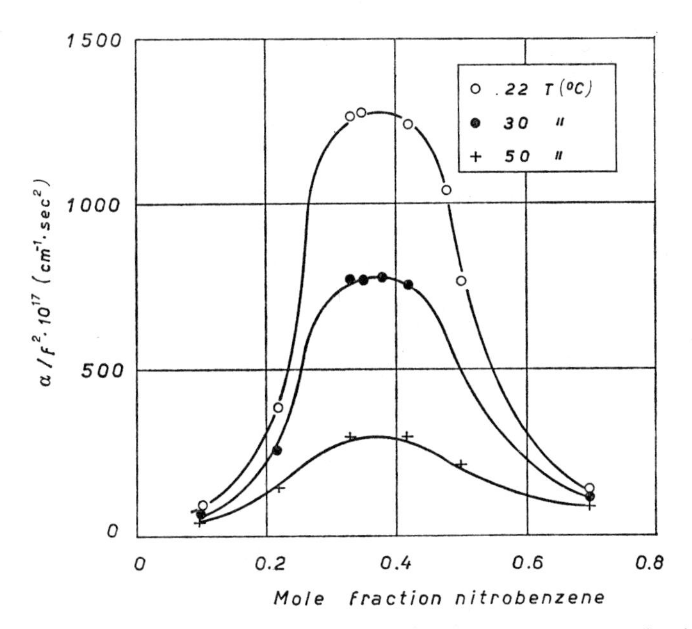

FIGURE 41. Sound absorption vs composition at same temperatures in nitrobenzene-*n*-hexane mixtures ($f = 15$ MHz) (D'Arrigo and Sette, 1968).

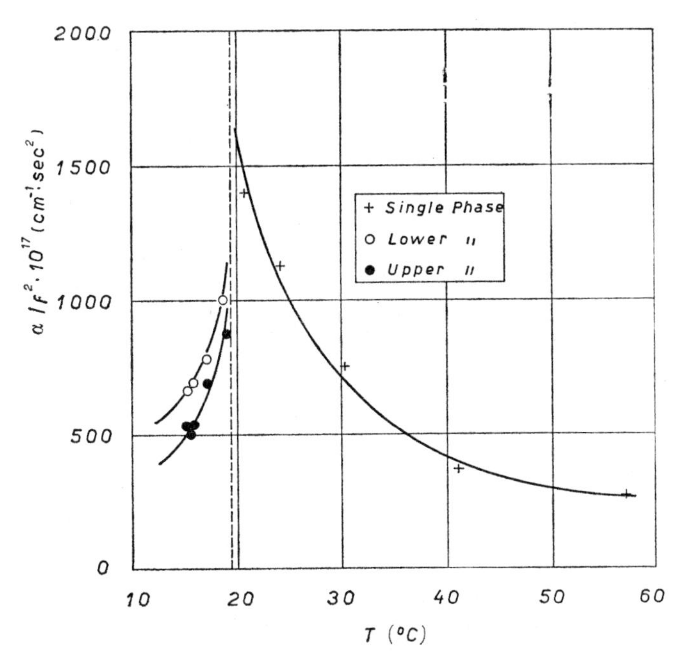

FIGURE 42. Sound absorption vs temperature for the 0.33 nitrobenzene mole fraction mixture of the system nitrobenzene-*n*-hexane (D'Arrigo and Sette, 1968).

critical temperature. Dispersion in critical mixtures has also been detected and will be mentioned later.

E. LIGHT SCATTERING

In Section 2.C we have seen how the total intensity of the light scattered through a wave vector **k** is related to the corresponding Fourier transform of the density-density correlation function. A structure factor $S(\mathbf{k})$ was introduced in the description of the scattering cross-section (equation 32). The dynamic information is contained in the spectrum of the scattered light.

The differential cross-section, i.e. the power per unit frequency (ω_s) interval scattered into a unit solid angle per unit scattering volume and per unit incident intensity

$$\frac{1}{V}\frac{\mathrm{d}^2\sigma}{\mathrm{d}\Omega\,\mathrm{d}\omega_s} = \frac{I_s(\omega)R^2}{VI_0{}^*} \tag{141}$$

can be expressed by means of a dynamic structure factor $S(\mathbf{k}, \omega)$

$$\frac{1}{V}\frac{\mathrm{d}^2\sigma}{\mathrm{d}\Omega\,\mathrm{d}\omega} = |A^2|\,\rho S\,|\mathbf{k}, \omega| \tag{142}$$

where

$$S(\mathbf{k}, \omega) = \frac{1}{2\pi}\int_V \mathrm{d}\mathbf{r}\int_{-\infty}^{\infty} \exp\,[i(\omega t + \mathbf{k}\cdot\mathbf{r})][G(\mathbf{r}, t) - \rho]\,\mathrm{d}t \tag{143}$$

is the Fourier transform in space and time of a space-time correlation function. $G(\mathbf{r}, t)$ is the generalization into the time domain of the radial distribution function† and it is the correlation between a particle at the origin at time $t = 0$ and one at **r** at time t. The function $S(\mathbf{k}, \omega)$ when integrated over frequency between $-\infty$ and $+\infty$ gives $S(\mathbf{k})$ of relation (32).

Equation (142) shows how the experimental determination of the spectrum of scattered light can furnish information on the important microscopic quantity $S(\mathbf{k}, \omega)$, the Fourier transform of $G(\mathbf{r}, t)$.‡

It is important now to see how $S(\mathbf{k}, \omega)$ can be linked to basic properties of the fluid. There is no simple general expression for $G(\mathbf{r}, t)$, as for $g(r)$. In a single fluid *far* from the critical point, the assumption can be made that the dynamics of the long wavelength density fluctuations ($\lambda = \frac{2\pi}{k}$ appreciably larger than the correlation length ξ, i.e. $k\xi \ll 1$) are described by classical

† However $G(\mathbf{r}, t) \xrightarrow[r\to\infty]{} \rho$ rather than 1.

‡ Similar relations are found between the neutron scattering cross-section and the two-spin correlation function.

hydrodynamics. The form of the spectrum can then be calculated (Mountain, 1966a) and the corresponding expression is named after Landau-Placzek.

The result has the form

$$\frac{1}{V}\frac{d^2\sigma}{d\Omega\, d\omega} = \frac{\pi^2}{\lambda_0^{\,4}}\left(\rho\frac{\partial\varepsilon}{\partial\rho}\right)^2 kTK_T\sigma_k(\omega) \tag{144}$$

where $\dfrac{\pi}{\lambda_0^{\,2}}\left(\dfrac{\partial\varepsilon}{\partial\rho}\right)$ replaces the scattering amplitude of equation (142) and $\sigma_k(\omega)$ has the expression

$$\sigma_k(\omega) = \frac{C_p - C_v}{C_p}\left[\frac{\dfrac{\lambda k^2}{\pi\rho_0 C_p}}{\left(\dfrac{\lambda k^2}{\rho_0 C_p}\right)^2 + \omega^2}\right] + \frac{C_v}{C_p}\left[\frac{\dfrac{\Gamma k^2}{2\pi}}{(\Gamma k^2)^2 + (\omega + u_0 k)^2} + \frac{\dfrac{\Gamma k^2}{2\pi}}{(\Gamma k^2)^2 + (\omega - u_0 k)^2}\right] \tag{145}$$

u_0 being the zero frequency sound velocity,

$$\Gamma = \frac{1}{2\rho_0}\left[(\tfrac{4}{3}\eta + \eta_B) + (\gamma - 1)\frac{\lambda}{C_0}\right] \tag{146}$$

One has also

$$\sigma_k(\omega) = \frac{S(\mathbf{k}, \infty)}{S(\mathbf{k})} \tag{147}$$

According to (145) the spectrum of $\sigma_k(\omega)$ is formed by three Lorentzian lines (Fig. 43a). The first is centered at $\omega = 0$ and has a strength $\dfrac{C_p - C_v}{C_p}$ and half width $\Gamma_R = \left(\dfrac{\lambda}{\rho_0 C_p}\right)k^2$. It is the so called quasi-elastic or Rayleigh component and it is produced by entropy fluctuations at constant pressure as the thermodynamic analysis shows. Such fluctuations are purely diffusive. The other two lines constitute the Brillouin doublet; they are displaced symmetrically from the Rayleigh line by $\omega_B = \pm u_0 k$. Their combined strength is $\dfrac{C_v}{C_p}$ and their half width $\Gamma_B = \Gamma k^2$. Their origin is found in the

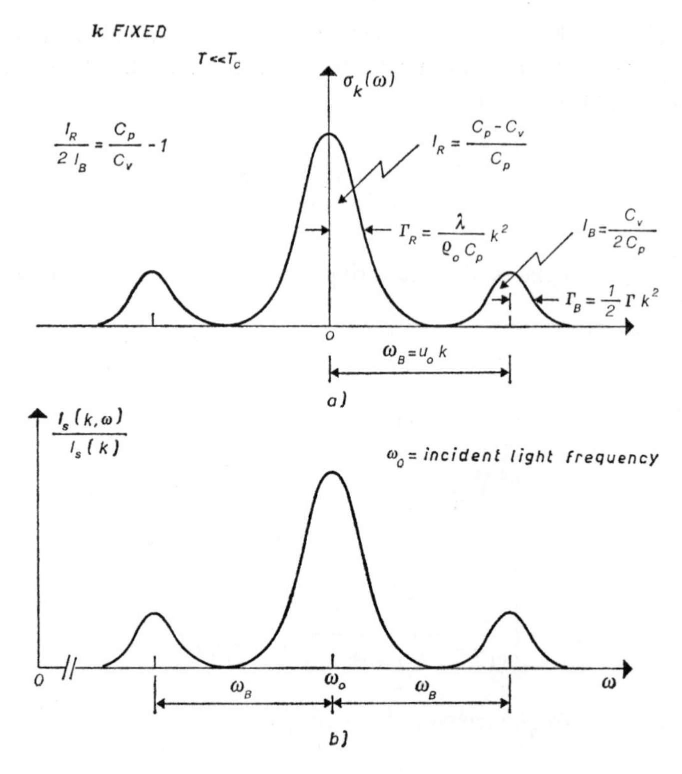

FIGURE 43. (a) Dependence of $\sigma_k(\omega)$ upon ω according to hydrodynamics for fixed T ($\ll T_c$), (b) Spectrum of scattered light at fixed $\bar{k}$ and $T \ll T_c$.

density fluctuation at constant entropy, i.e. in thermal sound waves in the fluid; the frequency displacement is due to Doppler effect and therefore depends on the sound velocity. The line-width instead is proportional to the sound absorption coefficient (equation 140). The spectrum of the scattered light has the same form as the spectrum of $\sigma_k(\omega)$ and it is given by the same curve simply shifted so that it is centered around the frequency (ω_0) of the incident light (Fig. 43b).

The separation of the spectrum in three lines occurs of course only when the Brillouin shift is much larger than the width of the lines. Otherwise the expression of the spectrum, as deduced from hydrodynamics is more complex. The spectrum is also more complicated when the particles have internal degrees of freedom (Mountain, 1966b).

Experiments conducted in many fluids far from the critical point with laser sources have shown the adequacy of the hydrodynamic treatment in these cases, and have furnished valuable information on many properties of the

Table 7

Quantities determined by light scattering in fluids in the hydrodynamic region $k\xi \ll 1$.

Quantity	Typical value in normal fluids
(1) Cross-section	
$\frac{1}{V}\frac{d\sigma}{d\Omega} = \frac{\pi^2}{\lambda_0^4}\left(\rho\frac{\partial\varepsilon}{\partial\rho}\right)^2 kTK_T$	10^{-6} cm^{-1}
(2) Landau-Placzek ratio	
$\frac{I_B}{2I_B} = \frac{C_p}{C_V} - 1$	1
(3) Rayleigh half linewidth	
$\Gamma_R = \frac{\lambda}{\rho_0 C_p} k^2$	$10^7 \frac{\text{rad}}{\text{sec}}$
(4) Brillouin shift	
$\omega_B = \mu_0 k$	$10^{11} \frac{\text{rad}}{\text{sec}}$
(5) Brillouin half linewidth	
$\Gamma_B = \Gamma k^2$	$10^9 \frac{\text{rad}}{\text{sec}}$
$\Gamma = \frac{1}{2\rho_0}\left[(\frac{4}{3}\eta + \eta_B) + (\gamma - 1)\frac{\lambda}{C_p}\right]$	

systems. Table 7 indicates the quantities which can be obtained in fluids (in the hydrodynamic region $k\xi \ll 1$) and their typical values. Recent developments of laser spectroscopy, based on the extension to light frequencies of the heterodyne and homodyne techniques used for high resolution detection of radiowaves (optical mixing spectroscopy) (Forrester *et al.*, 1947; Ford and Benedek, 1965; Lastovka and Benedek, 1966; Benedek, 1968, 1969; French *et al.*, 1969) have made possible accurate determination of the spectrum of scattered light (resolving powers of 10^{14} are reached).

In the critical region the condition $k\xi \ll 1$ is no longer valid and the dynamics of density fluctuations become much more complicated than in the classical hydrodynamic description. We may however consider how the various quantities should change as the critical temperature is approached while remaining in the range of validity of hydrodynamics.† The divergence of various quantities in $\sigma_k(\omega)$ determines special features of the spectrum.‡

(a) the Landau-Placzek ratio $\frac{I_R}{2I_B}$, i.e. the ratio of the light intensity in the Rayleigh line to that in the Brillouin lines, diverges as $\varepsilon^{-(\gamma-\alpha)}$; the Rayleigh component therefore becomes more and more predominant over the Brillouin lines as the critical point is approached.

(b) the Rayleigh linewidth approaches zero as $\varepsilon^{(\gamma-\psi)}$ if ψ is the critical exponent of the thermal conductivity ($\lambda \sim \varepsilon^{-\psi}$);

(c) the Brillouin shift approaches zero as $\varepsilon^{\alpha/2}$: the doublet moves in toward the central line as $T \to T_c$;

(d) the Brillouin linewidth depends through Γ on λ, η ζ, C_v and C_p and its behaviour as $T \to T_c$ depends on the behaviour and the relative importance of divergences of these parameters.

Let us give some experimental results starting from the linewidth of the Rayleigh line. The measurement of this linewidth, §

$$\Gamma_R = \chi k^2 = \frac{\lambda}{\rho C_p} k^2 = k^2 \frac{\lambda_0}{\rho C_{p0}} \frac{\varepsilon^{-\psi}}{\varepsilon^{-\gamma}} = k^2 \chi_0 \varepsilon^{\gamma-\psi} \tag{148}$$

as a function of temperature in pure fluids allows the determination of the temperature dependence of the thermal diffusivity χ. In the case of binary

† We remember that the inverse correlation range diverges

$$\kappa = \frac{1}{\xi} = \kappa_0 \varepsilon^{\nu}$$

and therefore, as $T \to T_c$ one always reaches a region where the relation $k\xi \ll 1$ is no longer valid. One may ask, however, how close to T_c experiments can be performed still remaining in the hydrodynamic region. In a simple fluid typical values of κ_0 are around 1 Å^{-1} and $\nu \simeq \frac{2}{3}$; therefore

$$\kappa \simeq \varepsilon^{2/3}(\text{Å}^{-1}) = 10^8 \varepsilon^{2/3}(\text{cm}^{-1})$$

If for instance experiments are performed with a helium-neon laser source ($\lambda = 6328$ Å) and the scattering is observed at 60°

$$k = \frac{4\pi}{\lambda} \sin \frac{\theta}{2} \simeq 10^{-3}\,\text{Å}^{-1} = 10^5\,\text{cm}^{-1}$$

One has $k \ll \kappa$ when $\varepsilon \gg 10^{-5}$. This means that the predictions of hydrodynamics should still be valid for a value of ε of about 10^{-3}–10^{-4}. The applicability of hydrodynamics can be extended somewhat by making measurements at smaller θ.

‡ Moreover as has been already mentioned (section II.C) the cross-section diverges as $K_T \sim \varepsilon^{-\gamma}$ (critical opalescence).

§ Standard thermodynamics gives the relation between the specific heat of a simple substance $C_p - C_v = TV\left(\frac{\partial p}{\partial T}\right)_V K_T$. Because C_v is positive and $\left(\frac{\partial p}{\partial T}\right)_V$ positive and finite, C_p has to diverge at the critical point as $K_T \sim \varepsilon^{-\gamma}$.

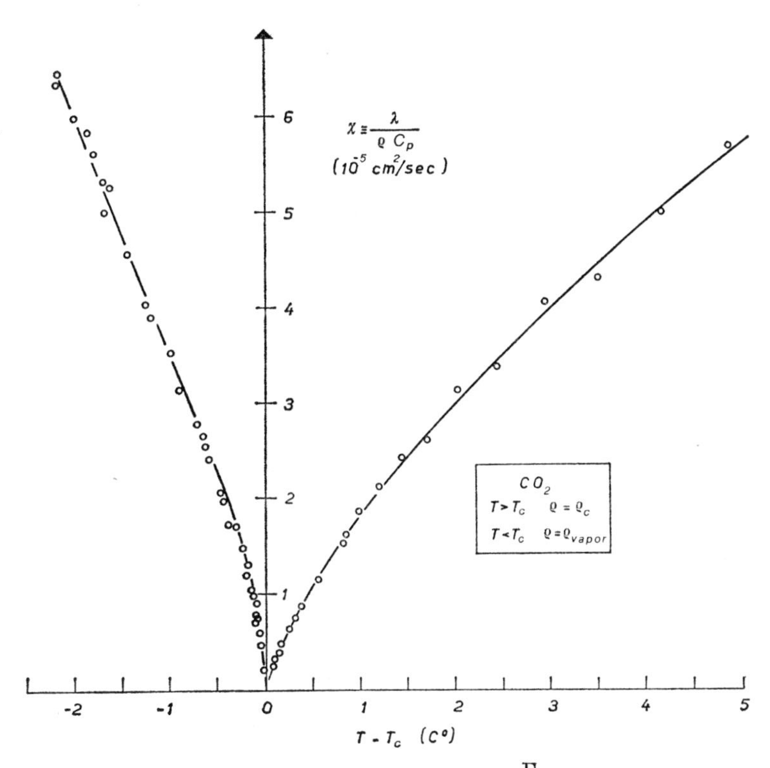

FIGURE 44. Thermal diffusivity $\chi = \frac{\Gamma_R}{k^2}$ of CO_2.

mixtures the analogous quantity is the diffusion constant D. Figure 44 (Swinney and Cummings, 1968) gives the thermal diffusivity $\chi = \frac{\Gamma_R}{k^2}$ as a function of temperature for CO_2 along the critical isochore ($T > T_c$) and along the vapour side of the coexistence curve ($T < T_c$).

The data so far obtained for $\chi(\varepsilon)$ and $D(\varepsilon)$ show differences in the critical exponents from one substance to another.† With the exception of SF_6 these data however indicate values of $(\gamma - \psi)$ in the range from 0.59 to 0.73 averaging $\sim \frac{2}{3}$. This value is in agreement with a strong divergence of λ already established in thermodynamic experiments. Moreover in almost all systems studied a reasonable symmetry above and below T_c has been found.

The light scattering experiments allow a check of the range of validity of the hydrodynamic treatment. According to this calculation in fact the log of

† Probably some of these differences are due to the fact that, as Sengers has pointed out, the experiment should be interpreted in terms of $\Delta\lambda$ and not of λ by taking into account background effects.

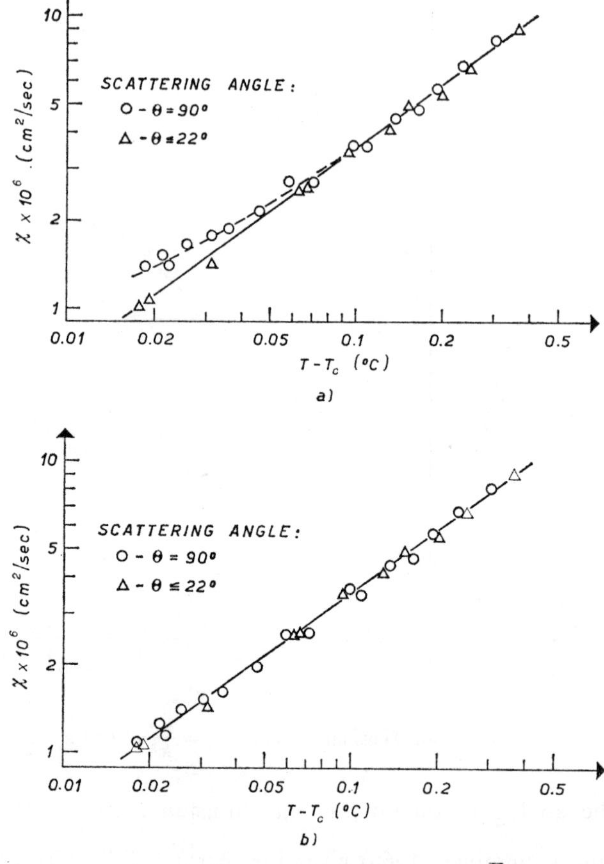

FIGURE 45. (a) Experimental data on thermal diffusivity $\chi = \frac{\Gamma_R}{k^2}$ in CO_2 (Swinney and Cummins, 1968). (b) Experimental data on thermal diffusivity and Rayleigh linewidth in CO_2 and Botch-Fixman relation (solid line) (Swinney and Cummins, 1968).

$\chi = \frac{\lambda}{\rho C_p}$ against the log of ε should be a straight line with slope given by $(\gamma - \psi)$. Figure 45a gives the Swinney and Cummins (1968) data on CO_2 along the critical isochore.

The hydrodynamic result is found valid for small scattering angle ($\theta = 22°$), while for large angle ($\theta = 90°$) the power law behaviour given by hydrodynamics fails when $T - T_c < 0.1°C$ ($\varepsilon \simeq 3 \cdot 10^{-4}$). Figure 45b shows how a modified expression (solid line) of the form

$$\Gamma_R = \chi k^2(1 + k^2\xi^2) \tag{149}$$

gives a better fit with experiments. This expression was suggested by Botch and Fixman (1965) and reduces to the Landau-Placzek expression when $k\xi \to 0$. The correction introduced by Fixman-Botch takes into account the increase in the correlation among particles as the critical point is approached. It is to be noted that analysis of the data by means of equation (149) enables the determination of $\xi(T)$. These data have been given in Fig. 25 (Cummins, 1970) for single fluids and binary mixtures (symbols not containing vertical

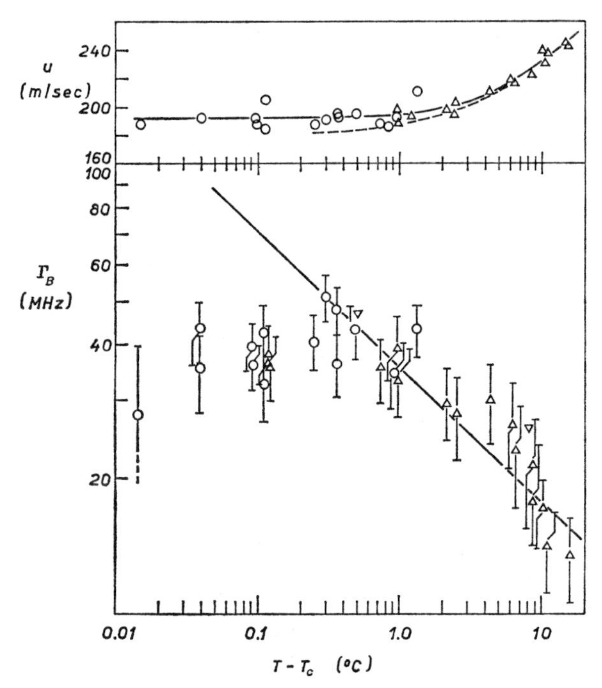

FIGURE 46. (a) Sound velocity in CO_2 deduced by Brillouin line shift (data of Ford *et al.*, 1968). (b) Half width of the Brillouin lines [Ford *et al.*, 1968 and (triangle) of Benedek and Cannell, 1968].

lines). The comparison of the value of ξ obtained for the same substance with the present technique and by means of the cross-section determination (Section 2.C) shows that the linewidth data are systematically smaller and suggests that probably the correction term $k^2\xi^2$ should be multiplied by a constant smaller than unity. Such a result is in agreement with theories which are discussed later.

Figure 46 gives a value for sound velocity deduced from the position of the Brillouin doublet in CO_2 and the halfwidth of the corresponding line. The data are from Ford *et al.* (1968) except for the triangles in part b which are from Benedek and Cannell (1968). Figure 47 gives the Landau-Placzek ratio along

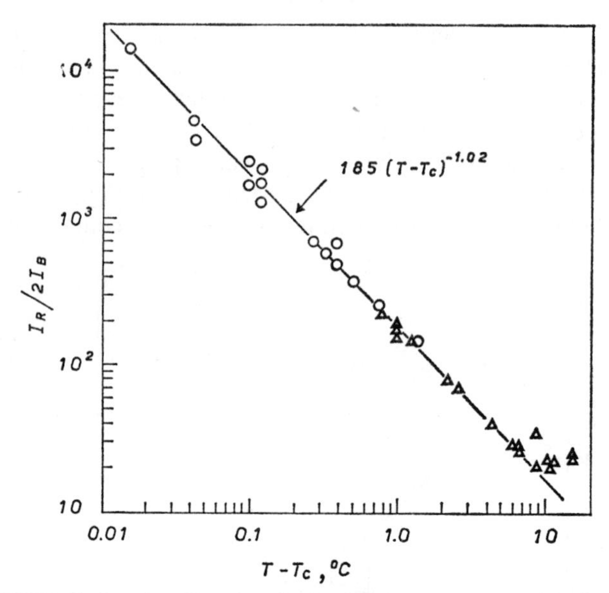

FIGURE 47. Landau-Placzek ratio for CO_2 vs temperature (Ford *et al.*, 1968).

the critical isochore in CO_2 as $T \to T_c^+$ (Ford *et al.*, 1968). This ratio diverges with an exponent 1.02, which is not very far from the hydrodynamic result $(\lambda - \alpha)$.

F. DYNAMIC SCALING LAWS

The efforts so far made to try to reach a satisfactory understanding of dynamic critical phenomena have followed essentially two lines and both have been fairly successful although neither is able, either to explain the entire body of experiments, or to give clear cut explanations of their meaning and limits. The first approach is an attempt to extend scaling ideas to dynamic processes; it is also capable of predictions beyond the validity of hydrodynamics. The second approach (mode-mode coupling theory) tries to reach a microscopic description of the interactions among different modes of excitation of the system (heat, viscous, sound modes) in order to calculate the divergent part of the transport coefficient with the help of static scaling laws; its validity is confined to hydrodynamic regimes.

As has been mentioned, at the present stage a satisfactory theory has not been found but the two above-mentioned approaches give results which mutually help to further the limited insight that each one proposes.

The dynamic scaling idea was first introduced by Ferrel *et al.* (1967) in the study of the λ-transition of helium and has been successively extended to other transitions by various authors. Outstanding contributions to the development of the theory have been given by Halperin and Hohenberg

(1969). We will consider their approach which directs attention to the correlation functions of densities of conserved quantities and quasiconstants of the motion in the equilibrium ensemble. These functions, which contain all the information relative to the macroscopic description (hydrodynamics) in the long-wavelength low frequency limit, may also be used outside the hydrodynamic domain.

In static scaling it is assumed that the order-parameter (ψ) correlation length $\xi(T)$ contains all the important effects concerning fluctuations so that it is the only important macroscopic length near T_c; it diverges at $T = T_c$. The behaviour of correlation functions depends essentially on the ratio between the wavelength $\lambda = \dfrac{2\pi}{k}$ and $\xi = \dfrac{1}{\kappa}$. The graph of Fig. 48 may be used to see more clearly the significance of the correlation length at various wave numbers, k. The origin of ξ^{-1} corresponds to T_c; on the right there is the region $T > T_c$, which corresponds to the disordered phase, and on the left $(T < T_c)$ is the ordered phase. The diagram is drawn under the assumption $k^{-1} \gg a$, $\xi \gg a$, a being a typical microscopic length such as the interparticle average spacing. Three asymptotic regions are identified, in each of which the correlation function of the order parameter $C^{\psi}(\mathbf{r})$ and its Fourier transform has a different characteristic behaviour. The regions I and III ($|k\xi| \ll 1$) correspond to phenomena which occur over a distance r large compared to ξ and can be described by macroscopic (hydrodynamic) equations; region II ($k\xi \gg 1$) is the real critical region where for both $T \gtrless T_c$ the phenomena

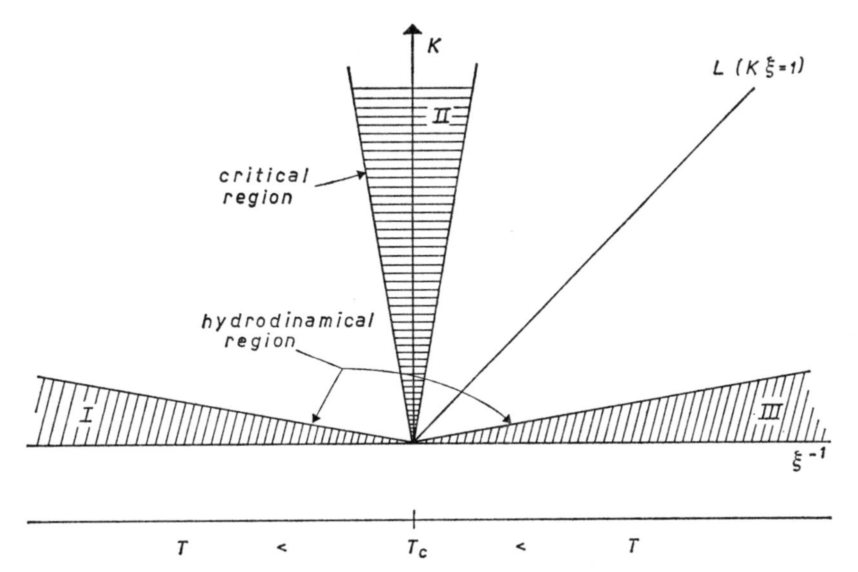

FIGURE 48. Regions of different dynamical behaviour.

occur over distances small compared to ξ, but large compared to all other relevant lengths. The Fourier transform of the static correlation function $C^{\psi}(\mathbf{k}, \xi)$ is strongly divergent at the origin of the diagram ($\mathbf{k} = 0$, $\xi^{-1} = 0$), but remains finite for any other $\mathbf{k}$ at $T = T_c$. The static scaling procedure is based on the hypothesis that the function $C^{\psi}(\mathbf{k}, \xi)$ varies smoothly in the plane k, ξ^{-1} except for the singularity at the origin. It can therefore be determined by its limiting behaviour in the three characteristic regions: this means that the forms valid in regions I and II extrapolated on a line such as L, ($k\xi = 1$), in Fig. 48 may differ by a factor of the order of unity. This means also that a single function of $k\xi$ describes the correlations in the whole (k, ξ^{-1}) plane, i.e. $C^{\psi}(\mathbf{k}, \xi)$ is an homogeneous function

$$C^{\psi}(\mathbf{k}, \xi) = k^{x} g(k\xi) \tag{150}$$

In an analogous way the dynamic scaling considers the dynamic correlation function of the order parameter and its Fourier transform $C^{\psi}(\xi, \mathbf{k}, \omega)$ as well as the normal mode which dominates its frequency spectrum. Such a mode may be diffusive or propagating and has a characteristic frequency $\omega^{\psi}(\xi, \mathbf{k})$. The dynamic scaling hypothesis assumes that the Fourier transform of the dynamic correlation function of the order parameter as well as $\omega^{\psi}(\xi, \mathbf{k})$, is completely determined by its behaviour in the three limiting regions. More precisely one may write

$$C^{\psi}(\xi, \mathbf{k}, \omega) = \frac{2\pi C^{\psi}(\xi, \mathbf{k})}{\omega^{\psi}(\xi, \mathbf{k})} \cdot f_k^{\psi}\left(\frac{\omega}{\omega^{\psi}}\right) \tag{151}$$

where f_k^{ψ} is a normalized shape function. The static scaling hypothesis involves a characteristic length ξ (or inverse length κ) and can be formulated assuming that $C_k^{\psi}(\xi, k)$ is a homogeneous function of k and ξ^{-1} (150). The dynamic scaling hypothesis involves a characteristic length (ξ) and a characteristic frequency ω^{ψ}† (or inverse time) and can be formulated by assuming:

(1) $\omega^{\omega}(k, \xi)$ is a homogeneous function of k and ξ^{-1}

$$\omega^{\psi}(k, \xi) = k^{z}\Omega(k\xi) \tag{154}$$

† A formal definition of $\omega^{\psi}(\xi, \mathbf{k})$ is obtained by requiring that precisely *half* of the total integrated area under a plot of the normalized shape function as a function of frequency arises from frequencies in the interval between $-\omega^{\psi}$, $+\omega^{\psi}$

$$\int_{-1}^{1} f_k^{\psi}\left(\frac{\omega}{\omega^{\psi}}\right) \mathrm{d}\left(\frac{\omega}{\omega^{\psi}}\right) = \frac{1}{2} \tag{152}$$

One has also

$$\int_{-\infty}^{\infty} f_k^{\psi}\left(\frac{\omega}{\omega^{\psi}}\right) \mathrm{d}\left(\frac{\omega}{\omega^{\psi}}\right) = 1 \tag{153}$$

(2) the normalized shape function $f_k^{\psi}\left(\frac{\omega}{\omega^{\psi}}\right)$ depends on k and T only through the product $k\xi$ (and the sign of ε).

The homogeneous functions of interest here, as for static scaling, can be determined by their limiting behaviour in the asymptotic region of Fig. 48; in particular in the hydrodynamic regions where they can be deduced from hydrodynamic theory.

All these scaling considerations have been presented for the correlation function of the order parameter of the phase transition (the density, for the fluid system); this scaling is called *restricted dynamic scaling*. The same procedure is frequently valid for correlation functions of other microscopic variables (as momentum density or energy density), the scaling in these cases is called *extended dynamic scaling*.

There exist however some correlation functions whose critical behaviour cannot be treated using the dynamic scaling hypothesis.

G. APPLICATIONS OF DYNAMIC SCALING

Let us now see some results of the application of dynamic scaling. As an example we shall consider the case of a normal simple fluid. The order parameter correlation function is, for this case, the density-density correlation function $G(\mathbf{r}, t)$ and its Fourier transform $[C^{\psi}(\xi, \mathbf{k})]$ is the structure factor given by equation (143). Relation (151) is

$$S(\xi, \mathbf{k}, \omega) = \frac{2\pi}{\omega^{\psi}(\xi, \mathbf{k})} S(\xi, \mathbf{k}) f_k\left(\frac{\omega}{\omega^{\psi}}\right) \tag{155}$$

In order to obtain a prediction from scaling one has to calculate f and ω^{ψ} for some region in the (ξ^{-1}, k) plane. This is possible in the hydrodynamic region ($k\xi \ll 1$) where we already have the expression for $S(\xi, k, \omega)$ given by (142), (144), and (145). Sufficiently near to T_c, in the region $T > T_c$, the Brillouin doublet disappears, the structure factor is dominated by the Rayleigh peak and its width (Γ_R) is just ω^{ψ}. Therefore putting $\frac{\omega}{\omega^{\psi}} = x$

$$\omega^{\psi}(\xi, \mathbf{k}) = \Gamma_R = \frac{\lambda}{\rho C_p} k^2 \tag{156}$$

$$f(\xi, \mathbf{k}, x) = \frac{1}{\pi} \frac{1}{x^2 + 1}. \tag{157}$$

using the critical exponents for C_p, λ, and since $\xi \sim \varepsilon^{-\nu}$ from (156)

$$\omega^{\psi}(\xi, \mathbf{k}) \sim \varepsilon^{\gamma-\psi} k^2 \sim \xi^{-(\gamma-\psi)/\nu} k^2 = k^{2+(\gamma-\psi)/\nu} (k\xi)^{-(\gamma-\psi)/\nu}. \tag{158}$$

The dynamical scaling hypothesis or the homogeneity of ω^{ψ} (154) then means

$$z = 2 + \frac{(\gamma - \psi)}{\nu}. \tag{159}$$

The same reasoning can be used in the hydrodynamic region for $T < T_c$, obtaining

$$z = 2 + \frac{(\gamma' - \psi')}{\nu'}. \tag{160}$$

If we assume the validity of the static scaling results $\gamma = \gamma'$, $\nu = \nu'$ one finds

$$\psi = \psi'. \tag{161}$$

The dynamic scaling indicates therefore symmetry of the thermal conductivity exponent above and below T_c. with a similar procedure one finds also the result that the form of the Fixman-Botch correction for the Rayleigh line-width expression (149) is correct.

Predictions on position, intensity and width of the Brillouin components of the scattered light can not be obtained from restricted scaling, because the entire weight of the spectrum near T_c is in the Rayleigh line. Predictions of this type can be obtained by extended scaling applied to momentum density-momentum density correlation functions. One finds that the sound wave damping constant D_s $(= 2\Gamma)$ should vary as

$$D_s \sim u_0\xi \sim u_0\xi_0\varepsilon^{-\nu} \tag{162}$$

and the Brillouin linewidth $\Gamma_B = \frac{1}{2}D_s k^2$ should diverge as the correlation length. This result does not agree with some microscopic calculations of the mode-mode coupling theory (Kadanoff and Swift, 1968) and with some experiments in CO_2 and Xe.

Dynamic scaling has been applied also to magnetic system by many authors. Inelastic scattering of neutrons is in many ways similar to light scattering (except for the different wavelength) and the dynamic structure factor has a similar expression in the two cases (Marshall and Lovesey, 1971). The experimental results seem to support the dynamic scaling prediction. Figures 49 and 50 show some results obtained by Collins *et al.* (1968) on the measurement of neutron diffraction in Fe, together with the lines predicted by the dynamic scaling. Figure 49 gives the temperature dependence of the spin wave stiffness constant D. The experimental data can be fitted with a line of slope 0.37 ± 0.03 which is close to the prediction. Figure 50 gives the k-dependence of the linewidth at T_c. The scaling prediction is a straight line with slope 2.5. The experiment gives the value 2.7.

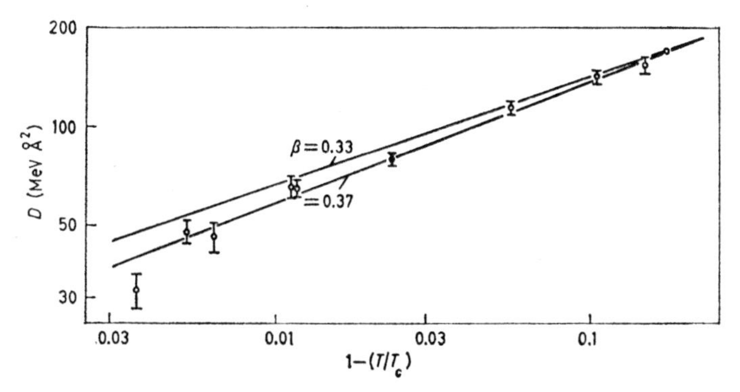

FIGURE 49. Spin wave stiffness constant in Fe near T_c (Collins *et al.*, 1969).

The first application of scaling ideas, as already mentioned, was proposed by Ferrell *et al.* (1967) in the study of the λ-transition of He; many other authors have considered the dynamic scaling law in this transition (see e.g. Halperin and Hohenberg, 1969). For $T < T_\lambda$ second sound dominates the spectrum of the order parameter fluctuations close to the transition and it constitutes therefore the critical mode. Scaling gives a relation between the

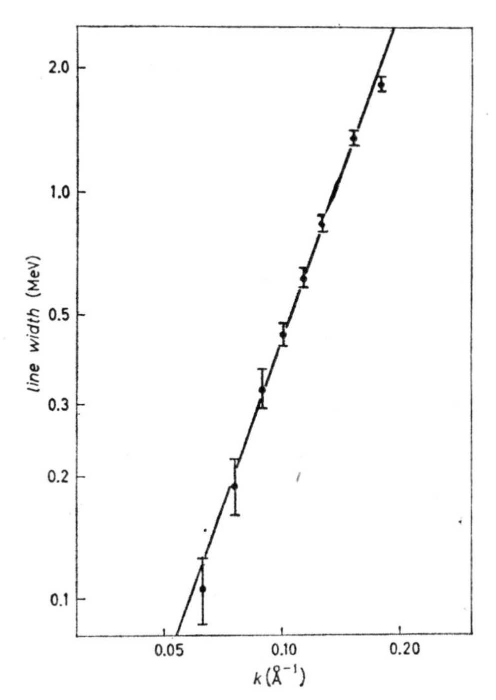

FIGURE 50. k-dependence of linewidth in Fe: $T = T_c$ slope 2.7 (Collins *et al.*, 1969).

decay rate of second sound (D_2), the second sound velocity (u_2) and ξ:

$$\frac{D_2}{u_2} = b\xi \tag{163}$$

b being a constant of the order of unity. This relation has been experimentally tested by Tyson (1968); the experimental results for D_2 plotted vs ($T - T_c$) (Fig. 51) lie along a straight line with slope ($-\frac{1}{3}$) as predicted by the theory. Ferrell *et al.* applied dynamic scaling to the thermal conductivity in He I: Ahlers (1968) results (Fig. 52) near T_λ in the single phase confirm the divergence predicted by scaling. The preceding discussion shows how many

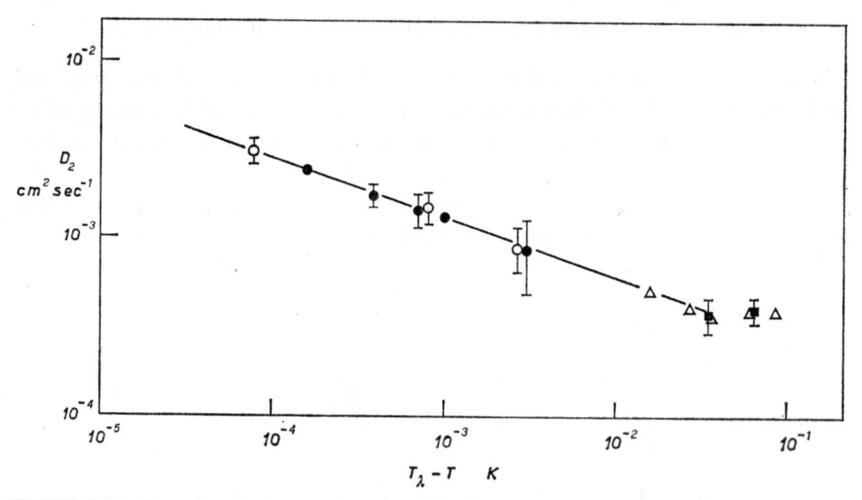

FIGURE 51. Damping constant of second sound in He II vs temperature near T_λ (Tyson, 1968).

experimental results give support to the dynamic scaling idea, as also for static scaling. It is to be stressed, however, that they do not give a proof of correctness, they only add to the confidence in the predictions of dynamic scaling.

The dynamic scaling procedure however has limitations; it is almost entirely based on the assumption of a single characteristic complex frequency $\omega(\mathbf{k})$ which practically determines the dynamic behaviour near T_c and which is a homogeneous function of k, vanishing at T_c. As Kadanoff and Swift have pointed out such a simple approach cannot be directly used when there is more than one characteristic frequency of different magnitude. The critical dynamics of the system in this case is determined by a complex mixing. In the case of pure fluids, for instance, three characteristic processes with different characteristic frequencies may be important and interact among themselves;

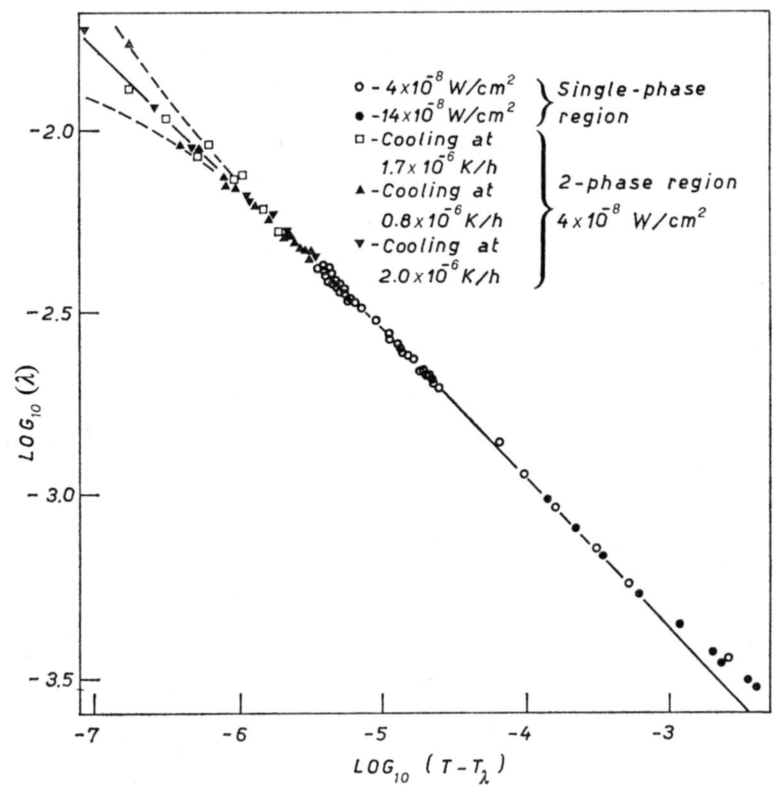

FIGURE 52. Thermal conductivity (λ) of liquid He I vs temperature near T_λ

the characteristic frequencies correspond respectively to the thermal relaxation rate, the viscous relaxation rate and the sound frequency at a wave vector of the order of the inverse correlation length.

H. MICROSCOPIC THEORIES. FIXMAN THEORY

The interest in dynamic scaling lies in the success of a large number of its predictions, although frequently these predictions cannot be pushed to describe completely the behaviour of a quantity. It has, for instance, been possible to predict the symmetry of the thermal conductivity exponent but no prediction of its value is given. As we have mentioned, another approach to dynamic critical phenomena has been initiated by a close examination of the microscopic processes which may occur in the system and may determine the observed behaviour of macroscopic quantities. This approach has led to treatments which although far from being conclusive, have had noticeable success in connecting the critical exponents of transport coefficients to the exponents of the static functions and in giving predictions often supported by experiments. A first attempt at a microscopic theory was given by Fixman (1962) in a calculation of viscosity in binary critical mixtures. In such a system near the critical point the composition fluctuations become very large especially in the long wavelength part of its spectrum. The leading idea of the Fixman approach is that in such a condition a velocity gradient introduced at the boundary for a viscosity experiment can easily induce concentration inhomogeneities; the return to homogeneity occurs through diffusion processes and leads to energy dissipation, which in a macroscopic description of the experiment results in an anomalous viscosity coefficient. This kind of approach has been applied by Fixman and co-workers [see Fixman (1964)] also to heat capacity and to sound velocity and absorption in fluids.

As an example let us consider in more detail the case of sound propagation in a binary critical mixture. According to Fixman the long-wavelength spatial fluctuations of composition give rise to an anomalous entropy and a complex dynamic heat capacity; the temperature variation produced by sound waves leads to relaxation processes and, as a consequence, to absorption and dispersion. The entropy fluctuations are expressed by means of integrals of the Fourier components of the radial distribution function; these integrals are extended between wave number zero and an upper cutoff in order to limit the considerations to long range correlation only. The Ornstein-Zernicke form of the correlation function is used and it is assumed that the dynamics of its long range part satisfy a diffusion equation. The results for sound velocity and for the absorption coefficient are:

$$\frac{u - \bar{u}}{\bar{u}} = -\tfrac{1}{2}H \,\mathrm{Re}\,[\varphi(d)] \tag{164}$$

$$\frac{\alpha}{f^2} = \frac{\pi H}{uf}\,\mathrm{Im}\,[\varphi(d)] \tag{165}$$

where $\bar{u}$ is the sound velocity in the absence of critical composition fluctuations, i.e. at frequencies sufficiently high that any velocity variation due to the process has disappeared. Moreover:

$$d = \kappa^2\left(\frac{h}{\omega}\right)^{1/2} \tag{166}$$

$$H = \frac{\gamma_0 - 1}{4\pi}\,\frac{R}{C_p{}^0}\left(\frac{\partial\kappa^2}{\partial\log T}\right)^2(n_1 + n_2)^2\left(\frac{h}{\omega}\right)^{1/4} \tag{167}$$

$$h = \frac{k_B T}{m_1}\,\phi_1 V_1^{\,2}(n_1 + n_2)(2\pi A\beta c_2) \tag{168}$$

$$\varphi(d) = \frac{1}{\sqrt{d}}\int_0^\infty \frac{x^4}{(x^2+1)\left[x^2(x^2+1) - \dfrac{1}{d^2}\right]}\,dx. \tag{169}$$

$C_p{}^0$ and γ_0 refer to heat capacities in the absence of critical composition fluctuations; A is one of the constants in the correlation function

$$g(r) = \frac{A}{r} \exp(-\kappa r);$$

m_i, n_i, c_i are molecular mass, number density and mass fraction of species i; ϕ_1 the volume fraction, V_i the partial molecular volume of species i; β is a friction constant which is related to the diffusion constant D_{id} of an equivalent ideal mixture $\beta = \dfrac{k_B T}{m_2 D_{id}}$; k_B the Boltzmann constant. Figure 53 gives the experimental results of D'Arrigo *et al.* (1970) for $\dfrac{\alpha}{f^2}$ in the critical mixture (44% mole fraction of aniline) of the system aniline-cyclohexane at $T - T_c = 0.7°C$ in the range 1–75 MHz as well as theoretical curves calculated with the Fixman theory, an approximation to it and with the theory for a single relaxation process. Fixman theory seems able to account fairly well for the experiments.

I. MODE-MODE COUPLING THEORIES

The basic idea proposed by Fixman may be summarized by saying that the presence of large long wavelength fluctuations near the critical point and the coupling among different transport modes may lead under external excitation to processes which, if the description of what occurs is still given by linearized equation of motion, correspond to anomalies in the transport coefficients. Such an idea has been used for a variety of calculations by many authors (Kawasaki, 1966, 1968a–d, 1969, 1970; Kawasaki and Tanaka, 1967; Deutch and Zwanzig, 1967; Mountain and Zwanzig, 1968; Villain, 1968; Ferrell, 1970; Kadanoff and Swift, 1968; Swift, 1968; Kadanoff, 1969c; Mistura, 1970).

Kadanoff and Swift (1968) have developed a special perturbation theory for the calculation of the rate of decay of hydrodynamic modes in order to determine transport coefficients near critical points in the gas–liquid system. In this theory they use static scaling laws to estimate correlation functions. This calculation has been extended to binary mixtures (Swift, 1968) and to the λ-transition of helium (Swift and Kadanoff, 1968). The transport modes are assumed to be non-linearly coupled to each other and the disturbance in the fluid passes from one mode to the other, back and forth. Figure 54 gives possible schemes. A perturbation calculation determines divergences in the transport coefficients near T_c. The origin of these divergences is in the slowly decaying intermediate states. In this theory those intermediate states which involve more than one transport process with long wavelength are considered in particular, i.e. processes by which a transport mode with wave

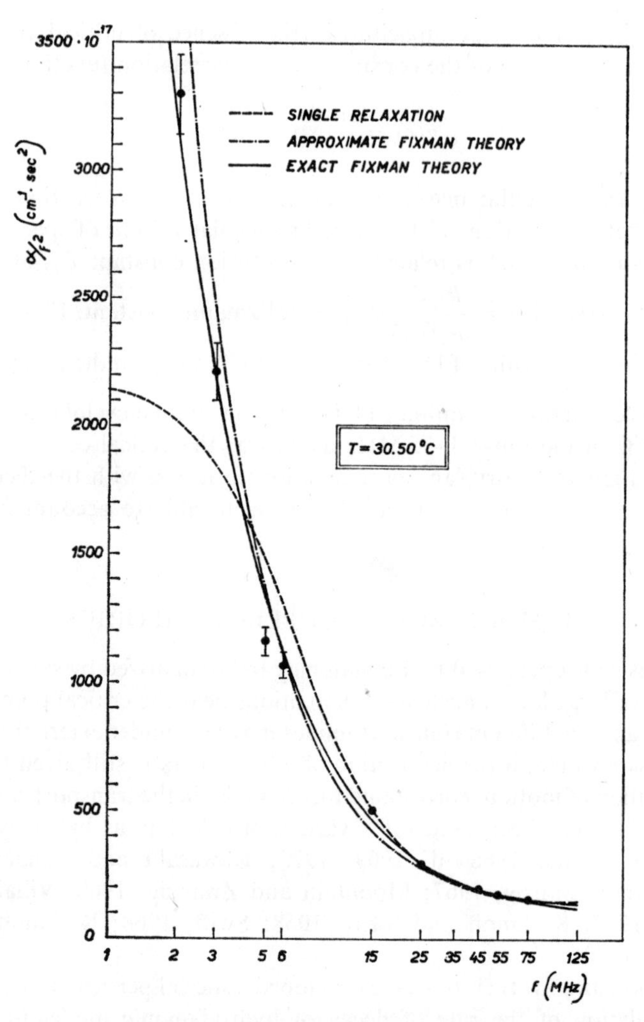

FIGURE 53. Sound absorption in aniline-cyclohexane critical mixture at $(T - T_c) =$ 0.7°C: experimental results and theoretical predictions (D'Arrigo *et al.*, 1970).

vector **k** decays by two independent processes with wave vectors **k**′ and (**k** − **k**′). These authors evaluate in this way contributions to heat conductivity from intermediate states involving a viscous mode and a heat mode or sound waves, as well as the contributions to the shear coefficient and to the longitudinal viscosity coefficient $(\frac{4}{3}\eta + \zeta)$ using schemes involving heat modes as intermediate states.

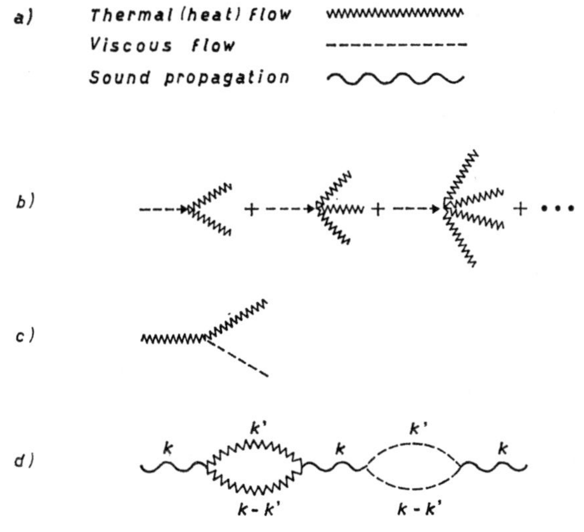

FIGURE 54. (a) Schematic representation of transport modes and sound propagation; (b) and (c) possible scheme of decay of (b) a viscous or of (c) a heat flow mode; (d) possible scheme of coupling of sound propagation with viscous or heat flow intermediate modes.

Three characteristic frequencies (better-inverse of relaxation time) are considered:

(1) $$s_T = \frac{\lambda}{\rho C_p}\,\xi^{-2} \tag{170}$$

which corresponds to the relaxation rate of the heat conduction mode with wave vector $k = \xi^{-1}$,

(2) $$s_\eta = \frac{\eta}{\rho}\,\xi^{-2} \tag{171}$$

which corresponds to the relaxation rate for transverse momentum flow at $k = \xi^{-1}$,

(3) $$s_s = u\xi^{-1} \tag{172}$$

which is the frequency of a typical sound wave with wavelength equal to ξ. These frequencies ($s_T < s_\eta < s_s$) delimit three regions.

The results of this calculation are given in Table 8 for the gas liquid system; in the table the relevant contribution to the various transport coefficients are

Table 8
Mode coupling calculations for transport coefficients.

	Region I	Region II	Region III	
	→ Increasing s →	→	→	→
	$s \sim s_T{}^* \sim \dfrac{\lambda^* \xi^{-2}}{\rho C_p} \sim \varepsilon^2$	$s \sim s_\eta{}^* \sim \dfrac{\eta^* \xi^2}{\rho} \sim \varepsilon^{4/3}$	$s \sim u\xi^{-1} \sim \varepsilon^{2/3}$	
Contributions to λ: from viscous flow plus heat modes	$\lambda \sim \dfrac{\rho C_p \xi^{-1}}{\beta \eta^*} \sim \varepsilon^{-2/3}$	→		
from sound waves plus heat modes	$\lambda \sim \dfrac{\xi^{-2} C_p}{u\beta} \sim \varepsilon^0$	→	→	
Contributions to η: from heat modes	$\eta \sim \dfrac{\rho C_p \xi^{-1}}{\beta \lambda^*} \sim \varepsilon^0$			
from sound waves plus heat modes	—	$\eta \sim \dfrac{C_p \xi^{-2}}{u\beta C_V} \sim \varepsilon^0$	→	
Contributions to ζ: from heat modes	$\zeta \sim \dfrac{\rho^2 u^2 C_p \xi^2}{\lambda^*} \sim \varepsilon^{-2}$			
from sound waves plus heat modes	—	$\zeta \sim \rho u \xi \sim \varepsilon^{-\nu+\alpha/2} \sim \varepsilon^{-2/3}$	→	
From high-k processes	—	$\eta \sim \zeta \sim \lambda \sim$ constant	→	→

* Are quantities at wave number ξ^{-1}

given as well as an evaluation of the temperature dependence of these coefficients. For such an evaluation it has been assumed that $\xi \sim \varepsilon^{-2/3}$, $C_p \sim \varepsilon^{-4/3}$ and that C_v diverges logarithmically.

A divergence is foreseen for the thermal conductivity on the critical isochore and the exponent predicted $(-\frac{2}{3})$ agrees with experiment (Section III,C on CO_2). The shear viscosity should be either strongly cusped at the critical point or weakly divergent. The same results were obtained by Swift (1968) in the extension of the theory to binary mixtures. The experiments in binary mixtures (Section III,B) agree with this result; the results in simple fluids have insufficient accuracy to determine the kind of anomaly present. The theory predicts a strong divergence for the bulk viscosity; it should diverge as ε^{-2} for low frequencies and as $\varepsilon^{-2/3}$ for higher frequencies at the critical isochore near the critical point. The same analysis shows that the sound wave damping constant becomes in region II and III

$$D_s = Au\xi \tag{173}$$

A being a constant of the order of unity. The Kadanoff and Swift calculation has been extended to binary mixtures, as already mentioned, by Swift (1968) and to the λ-transition of helium (Swift and Kadanoff, 1968). The theoretical predictions are usually in good agreement with experimental results.

The basic idea of Fixman has been used in another approach to mode-mode coupling in an extensive series of papers by Kawasaki. The results are largely coincident with those of Kadanoff and Swift. Table 9 gives the

Table 9
Critical exponents in Kadanoff-Swift and Kawasaki theories

	gas liquid			binary mixtures			
	λ	η	ζ	λ	η	ζ	D
Kadanoff Swift	$\varepsilon^{-\gamma+\nu}$	ln ε or finite	$\varepsilon^{-\gamma-\nu+\frac{3}{2}a}$	finite	ln ε or finite	ε^{-2}	ε^{ν}
Kawasaki	$\varepsilon^{-\gamma+\nu}$	finite	$\varepsilon^{2\alpha-2}$	—	finite	ε^{-2}	ε^{ν}

prediction for the transport properties of fluids furnished respectively by the Kadanoff and Swift (1968), and Swift (1968) calculations and by Kawasaki's (1970a, b) theory.

The Kawasaki calculations have given better predictions concerning the scattered light. A combination of the Kadanoff-Swift calculations and dynamic scaling had already given for the width of the Rayleigh component

$$\Gamma_R \xrightarrow[k\xi\to 0]{} \chi_0\xi_0 \frac{k^2}{\xi} \tag{174}$$

$$\Gamma_R \xrightarrow[k\xi\to\infty]{} Ak^3 \tag{175}$$

with a predicted temperature independent behaviour proportional to k^3 as $k\xi \to \infty$. Kawasaki has obtained the result

$$\Gamma_R = \frac{8A}{3\pi\xi^3} K(k\xi) \tag{176}$$

$K(x)$ being the so called Kawasaki function

$$K(x) = \frac{3}{4}\left[1 + x^2 + \left(x^3 - \frac{1}{x}\right) \tan^{-1} x\right] \tag{177}$$

and

$$A = \frac{k_B T}{16\eta_s} \tag{178}$$

In the two limits one has

$$\Gamma_R \xrightarrow[k\xi \to 0]{} \frac{8A}{3\pi\xi} k^2[1 + \tfrac{3}{5}(k\xi)^2 + \cdots] \tag{179}$$

$$\Gamma_R \xrightarrow[k\xi \to \infty]{} Ak^3 \tag{180}$$

Such a solution, which agrees with the Kadanoff and Swift calculations in both the hydrodynamic and critical limits gives however more information. In the first place if we compare (179) with (149) it appears that the Kawasaki correction to the Landau-Placzek relation for the Rayleigh line-width has the same form as the Fixman-Botch term; a multiplicative factor, $\frac{3}{5}$ is however included. A recent evaluation of the existing experimental data by Chu (1970) supports the Kawasaki result. The calculation gives also for the thermal diffusivity

$$\chi = \frac{8A}{3\pi\xi} = \frac{kT}{6\pi\xi\eta_s} \tag{181}$$

i.e. it is determined by ξ and η_s.

More important is the conclusion concerning the value of the constant A which can be experimentally tested. Berge *et al.* (1969, 1970) made measurements in the system aniline-cyclohexane and verified the prediction of a temperature independent limiting form and evaluated $A = 1.52\ 10^{13}$ cm³. According to the Kawasaki expression such a value of A indicates a value of

1.97 10^{-2} c.g.s. for η_s in remarkable agreement with experiment (η = 1.90 10^{-2} c.g.s. found by Arcovito *et al.*, 1969). Moreover the ratio $\frac{\Gamma_R}{Ak^3}$ should be 1 in the critical limit ($k\xi \gg 1$) and $\frac{8}{3\pi k\xi}$ in the hydrodynamic limit. A plot of $\log \Gamma_R/Ak^3$ versus $\log \frac{1}{k\xi}$ (Kawasaki plot) should have a flat intercept with the ordinate axis equal to 1 and asymptotic slope $\left(\text{as } \frac{1}{k\xi} \to \infty\right)$ of 1. Figure 55 gives the results of Berge *et al.* (1970) in the quoted mixture

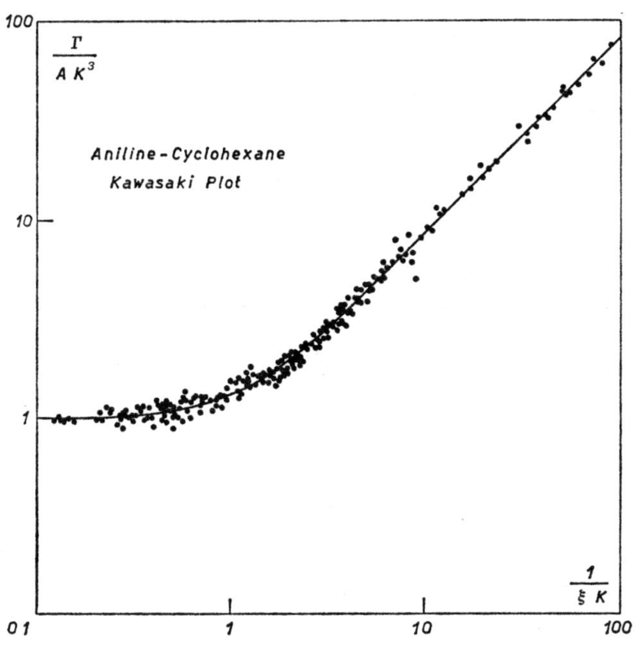

FIGURE 55. Kawasaki plot of Rayleigh line width data for aniline-cyclohexane (Berge *et al.*, 1970).

and Fig. 56 gives the analogous plot for Xenon (Henry *et al.*, quoted in Cummins, 1970).

Kawasaki has used his coupling theory for an extensive treatment of sound propagation in a critical gas-liquid system when the sound wavelength is much larger than the correlation length. He has considered the contribution to sound absorption coming from various couplings, i.e. through the coupling with two heat modes, two viscous modes, two sound waves, a viscous mode and a heat mode. He has found that throughout the frequency range of interest, $0 < f < \varepsilon^{\nu}$, the predominant contribution comes from heat intermediate modes.

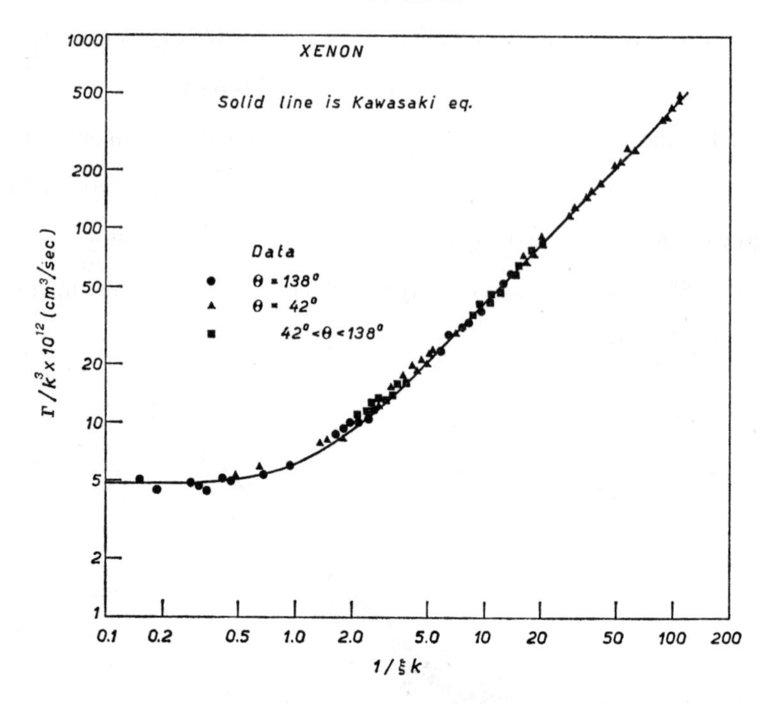

FIGURE 56. Kawasaki plot of Rayleigh line width data in Xenon (Henry *et al.*, quoted in Cummins, 1970).

The expressions for absorption coefficients and dispersion that he finds in the lowest of three frequency ranges ($0 < f \ll \varepsilon^{3\nu}$) are those which seem to give a fairly good comparison with experiments.

The formulae in this range have been obtained also from a simpler treatment which starts from the Fixman approach by Mistura (1970). Further developments are due to Tartaglia *et al.* (1972) and Eden *et al.* (1972). In agreement with the result that the process involves only heat modes as intermediate modes, it is found that experimental data at different frequencies suitably treated can be represented by one curve only, if a reduced frequency is used $f^* = \frac{f}{\bar{f}}$, where $\bar{f}$ is a characteristic frequency defined for simple fluid as

$$\bar{f} = \frac{\bar{\omega}}{2\pi} = \frac{\lambda}{\pi \rho C_p} \xi^{-2}. \tag{182}$$

It corresponds to the relaxation rate of the heat conduction mode with wave vector $k \simeq \xi^{-1}$. For binary mixtures where the mass diffusion mode takes the role played by the conduction mode in one component fluid, we have instead

$$\bar{f} = \frac{\bar{\omega}}{2\pi} = \frac{D}{\pi}\,\xi^{-2}. \tag{183}$$

Without going into more detail we present some experimental results in a simple fluid and in a mixture together with the curves that these theories yield for the relevant quantities.

Figures 57 and 58 refer to Xe (Eden *et al.*, 1972): the experimental data

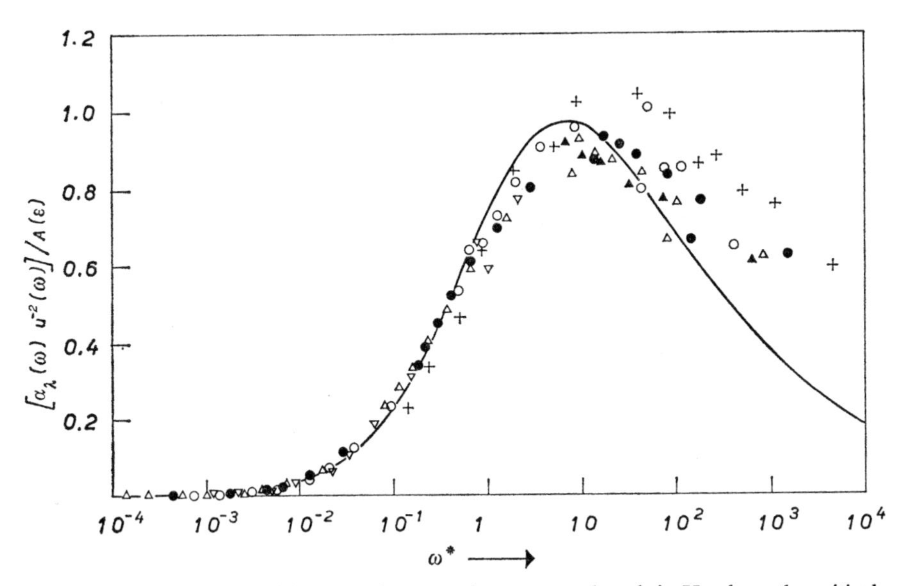

FIGURE 57. Reduced critical sound attenuation per wavelength in Xe along the critical isochore ($T > T_c$) as a function of the reduced frequency. Ultrasonic data at: 0.4 (solid triangle), 0.55 (open triangle); 1 (solid dot), 3 (open dot) and 5 (inverted triangle) MHz. Hypersonic data from Brillouin scattering at ~ 500 MHz and ~ 170 MHz (solid squares) (Eden *et al.*, 1972).

were obtained with ultrasonic techniques (0.5–5 MHz) and with Brillouin scattering [170 MHz, 500 MHz]; $A(\varepsilon)$ and $B(\varepsilon)$ are quantities weakly dependent on temperature. The solid lines are given by the theory. Although some discrepancies still exist, the agreement appears to be fairly good. Figures 59 and 60 give the results in the aniline-cyclohexane critical mixture for the absorption per wavelength (D'Arrigo *et al.*, 1971) and for the dispersion. In these figures the ordinate is reduced absorption per wavelength (α_λ*) and reduced sound velocity (u*) respectively. The solid lines are given by the theory. The agreement in this case also seems good.

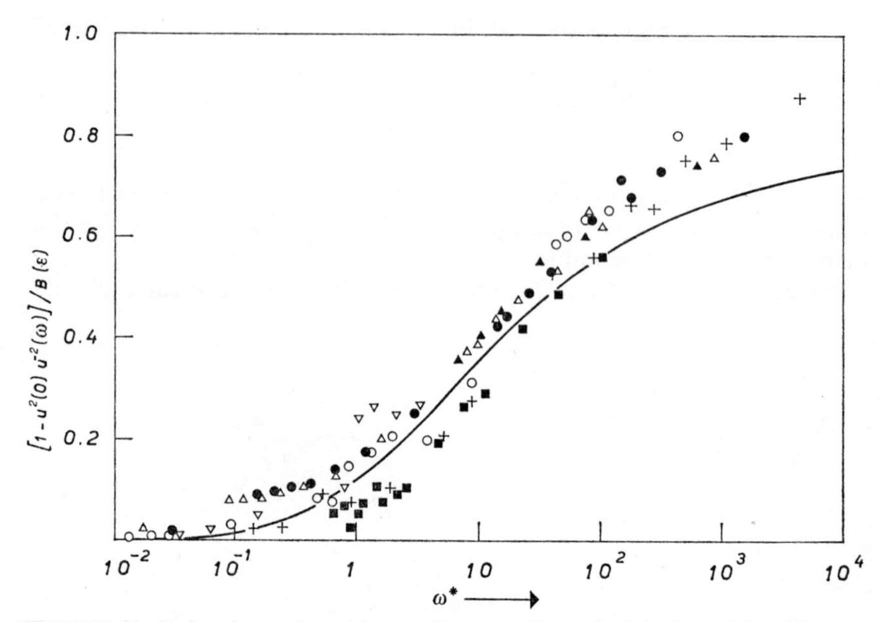

FIGURE 58. Reduced sound velocity on Xe along the critical isochore ($T > T_c$) as a function of the reduced frequency. Symbols for experimental data as in Fig. 57 (Eden *et al.*, 1972).

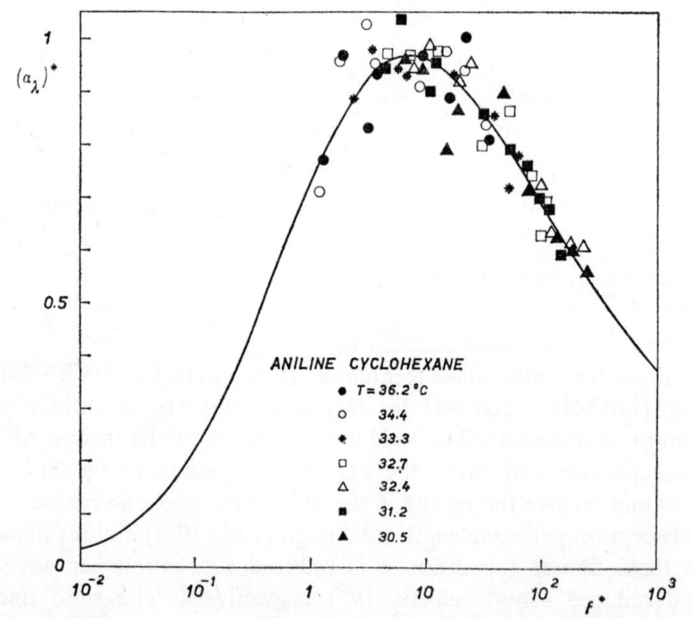

FIGURE 59. Reduced critical sound absorption per wavelength in the aniline cyclohexane critical mixture as a function of the reduced frequency f^* (D'Arrigo *et al.*, 1971).

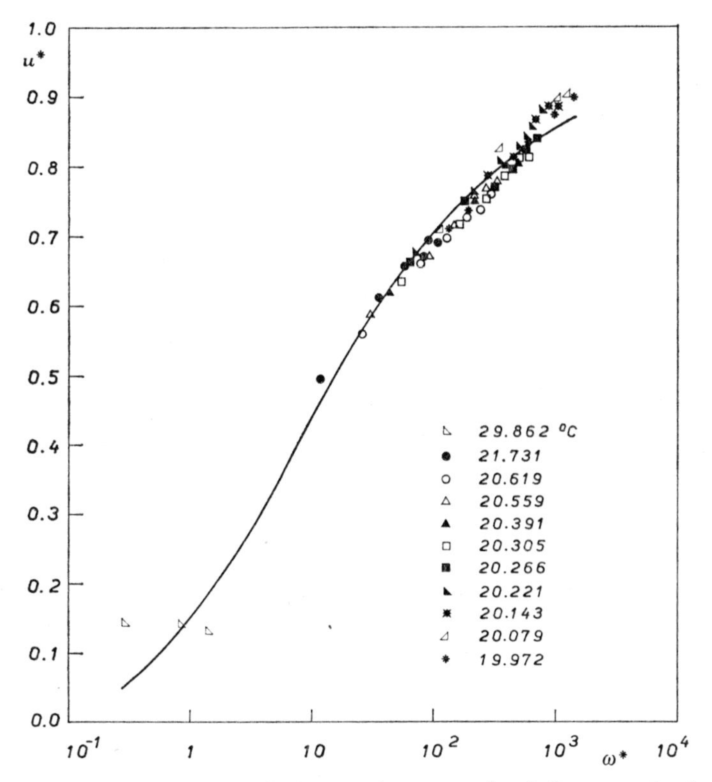

FIGURE 60. Reduced velocity of ultrasound versus reduced frequency in the nitrobenzene-*n*-hexane system (Tartaglia *et al.*, 1972).

J. SCALING AND MODE COUPLING

The preceding sections have merely indicated the different philosophies of the two approaches and the success that they have in making predictions which are found to agree with experiments. It is not our purpose to give a full coverage of the work done up to now on the various critical systems and of the various attempts that are presently being made to improve further these theories and our understanding of their foundations. We wish only to note how the two approaches we have illustrated when they are able to make predictions, usually give results which agree with each other.

Attempts are now being made to try to understand this circumstance and see if it is possible to establish a connection between the dynamic scaling and the mode-mode coupling theories. Kawasaki (1970) has started a study of the dynamics of critical fluctuations near T_c according to the mode-coupling approach. Near the critical point the description of the system requires more variables than the macroscopic ones (those for which the conservation principle holds, namely local density, local momentum density, local energy density). These new variables (hidden variables) describe the multiplicity of hydrodynamic modes which couple with a given hydrodynamic mode. Kawasaki has found that in certain cases it is possible to choose a set of dynamical variables (macroscopic variables plus hidden variables) whose dynamics are asymptotically closed and whose frequency spectrum can be predicted by dynamical scaling. This is a first slight indication of possible links between the two approaches.

Future research will surely clarify the exact nature of these links.

References

Ahlers, G. (1968). *Phys. Rev. Lett.* **21,** 1159.

Als-Nielsen, J. and Dietrich O. W. (1967). *Phys. Rev.* **153,** 706.

Andrews, T. (1869). *Phil. Trans. R. Soc.* **159,** 575.

Arcovito, G., Faloci, G., Mistura, L., and Roberti, M. (1969). *Phys. Rev. Lett.* **22,** 1040.

Baxter, R. J. (1972). *Ann. Phys.* (*N.Y.*) **70,** 193.

Begastskii, M. I., Voronel, A. V., and Gusak, B. G. (1962). *Zurn. Exsp. Teor. Fiz.* **43,** 728.

Begastskii, M. I., Voronel, A. V., and Gusak, B. G. (1963). *Sov. Phys. JEPT* **16,** 517.

Benedek, G. B. (1969). *In* Polarization, Matter and Radiation. Presse Universitaire de France, Paris.

Benedek, G. B. and Cannell, D. S. (1968). *Bull. Am. Phys. Soc.*, **13,** 182.

Berge, P., Calmettes, P., Laj, C., and Volochine, B. (1969). *Phys. Rev. Lett.* **23,** 693.

Berge, P., Calmetter, P., Laj, C., Tournarie, M. and Volochine, B. (1970). *Phys. Rev. Lett.* **24,** 1223.

Botch, W. and Fixman, M. (1965). *J. Chem. Phys.* **42,** 199.

Bragg, W. L. and Williams, E. J. (1934). *Proc. R. Soc.* **145,** 699.

Buckingham, M. J., Fairbank, W. M., and Kellers, C. F. (1961). *In* Progress in Low Temperature Physics, Vol. III. North-Holland, Amsterdam.

Buckingham, M. J. and Gunton, J. O. (1969). *Phys. Rev.* **178,** 848.

Chu, B., Kunahara N., and Fenby, D. V. (1970). *Phys. Lett. A* **32,** 131.

Chynoweth, A. G. and Schneider, W. G. (1952). *J. Chem. Phys.*, **20,** 1777.

Clow, J. R. and Reppy, J. D. (1966). *Phys. Rev. Lett.* **16,** 887.

Collins, M. F., Minkievicz, V. J., Nathans, R., Passell, L., and Shirane, G. (1969). *Phys. Rev.* **179,** 417.

Cummins, H. Z. (1970). *In* Proceedings of the International School of Physics "E. Fermi" on Critical Phenomena (ed. Green M. S.) Academic Press, London and New York.

D'Arrigo, G. and Sette, D. (1968). *J. Chem. Phys.* **48,** 691.

D'Arrigo, G., Mistura, L., and Tartaglia, P. (1970). *Phys. Rev.* **A1,** 286.

D'Arrigo, G., Mistura, L., and Tartaglia, P. (1971). *Phys. Rev.* **A3,** 1718.

Debye, P. (1959). *J. Chem. Phys.* **31,** 680.

De Pasquale, F., di Castro, G., and Jona-Lasinio, G. (1970). *In* Proceedings of the International School of Physics "E. Fermi" on Critical Phenomena (ed. Green, M. S.) Academic Press, London and New York.

De Pasquale, F. and Tombesi, P. (1972). *Nuovo Cimento Serie 11*, **12B**, 43.
Deutch, J. M. and Zwanzig, R. (1967). *J. Chem. Phys.* **46**, 1612.
Domb, C. (1960). *Adv. Phys.* **9**, 149.
Domb, C. and Hunter, D. L. (1965). *Proc. Phys. Soc.* **86**, 1147.
Eden, D., Garland, C. W., and Thoen, J. (1972). *Phys. Rev. Lett.* **25**, 726.
Egelstaff, P. A. and Ring, J. W. (1968). *In* Physics of Simple Liquids (eds. Temperley, Rowlinson; Rushbruke). North-Holland, Amsterdam.
Essam, J. W. and Fisher, M. E. (1963). *J. Chem. Phys.* **38**, 802.
Ferer, M., Moore, M. A. and Wortis, M. (1969). *Phys. Rev. Lett.* **22**, 1382.
Ferrell, R. A., Menyhard, N., Schmidt, H., Schwabl, F., and Szepfaluzy, P. (1967). *Phys. Rev. Lett.* **18**, 891; *Phys. Lett.* **24A**, 493.
Ferrell, R. A. (1970). *Phys. Rev. Lett.* **24**, 1169.
Fisher, M. E. (1964). *J. Math. Phys.* **5**, 944.
Fisher, M. E. (1967). Report of Progress in Physics 30, Part II, 615.
Fisher, M. E. (1969). *Phys. Rev.* **180**, 584.
Fixman, M. (1962). *J. Chem. Phys.* **36**, 310.
Fixman, M. (1964). Advances in Chem. Phys. vol. 6, ed. Prigogine I.
Ford, N. C. and Benedek, G. B. (1965). *Phys. Rev. Lett.* **15**, 649.
Ford, N. C., Langley, K. H., and Puglielli, V. G. (1968). *Phys. Rev. Lett.* **21**, 9.
Forrester, A., Parkins, W. E., and Grjnov, E. (1947). *Phys. Rev.* **72**, 728.
French, M. J., Angus, J. C., and Walton, A. G. (1969). *Science, N.Y.* **163**, 345.
Friendlander, J. (1901). *Z. Physik Chem.* **38**, 385.
Gammon, R. W., Swinney, H. L., and Cummings, H. Z. (1967). *Phys. Rev. Lett.* **19**, 1467.
Green, M. S., Sengers, J. V. (1966). Critical Phenomena—Proceedings of the Washington Conference (1965) Nazional Bureau of Standards Pubbl. 273.
Griffiths, R. B. (1965). *J. Chem. Phys.* **43**, 1958.
Griffiths, R. B. (1967). *Phys. Rev.* **158**, 176.
Guggenheim, E. A. (1945). *J. Chem. Phys.* **13**, 253.
Habgood, H. W. and Schneider, W. G. (1954). *Cand. J. Chem.* **32**, 98.
Halperin, B. J. and Hohenberg, P. C. (1969). *Phys. Rev.* **177**, 952.
Heller, P. (1967). Report of Progress in Physics 30, Part. II, 731.
Heller, P. and Benedek, G. B. (1962). *Phys. Rev. Lett.* **8**, 428.
Ho, J. T. and Lister, J. D. (1969). *Phys. Rev. Lett.* **22**, 603.
Kac, M., Uhlenbeck, G. E., and Hemmer, P. C. (1963). *J. Math. Phys.* **4**, 216.
Kadanoff, L. P. (1966). *Physics* **2**, 263.
Kadanoff, L. P. (1969a). *Phys. Rev.* **188**, 859.
Kadanoff, L. P. (1969b). *Phys. Rev. Lett.* **23**, 1430.
Kadanoff, L. P. (1969c). *J. Phys. Soc. Japan* **26**, 122.
Kadanoff, L. P. (1970). *In* Proceedings of the International School of Physics "E. Fermi" on Critical Phenomena (ed. Green, M. S.) Academic Press, London and New York.
Kadanoff, L. P. and Swift, J. (1968). *Phys. Rev.* **166**, 89.
Kadanoff, L. P., Gotze, W., Hamblen, D., Hecht, R., Lewis, E., Palciauskas, V. V., Rayl, M., Swift, J., Arpnes, D., Kane, J. (1967). *Rev. Mod. Phys.* **79**, 395.
Kao, W. P. and Chu, B. (1969). *J. Chem. Phys.* **50**, 3986.
Kawasaki, K. (1966). *Phys. Rev.* **150**, 291.
Kawasaki, K. (1968a). *Prog. Theoret. Phys.* (*Kyoto*) **39**, 1133.
Kawasaki, K. (1968b). *Prog. Theoret. Phys.* (*Kyoto*) **40**, 11.
Kawasaki, K. (1968c). *Prog. Theoret. Phys.* (*Kyoto*) **40**, 706.

Kawasaki, K. (1968d). *Prog. Theoret. Phys. (Kyoto)* **40,** 930.
Kawasaki, K. (1969). *Prog. Theoret. Phys. (Kyoto)* **41,** 1190.
Kawasaki, K. (1970a). *Ann. Phys. N.Y.* **61,** 1.
Kawasaki, K. (1970b). *Phys. Rev.* **A1,** 1750.
Kawasaki, K. (1970c). *In* Proceedings of the International School of Physics "E. Fermi" on Critical Phenomena (ed. Green, M. S.) Academic Press, London and New York.
Kawasaki, K. and Tanaka, M. (1967). *Proc. Phys. Soc.* (Lond.) **90,** 791.
Kouvell, J. S. and Conly, J. B. (1968). *Phys. Rev. Lett.* **20,** 1237.
Josephson, B. D. (1967). *Proc. Phys. Soc.* **92,** 269, 276.
Landau, L. D. (1937). *see* Statistical Physics by Landau, L. D. and Lifshitz, E. M. (1958). Pergamon Press, Oxford.
Landau, L. D. and Lifschitz, E. M. (1958). Statistical Physics. Pergamon Press, Oxford.
Lastovka, H. B. and Benedek, G. B. (1966). *Phys. Rev. Lett.* **17,** 1039.
Leister, H. M., Allegra, J. C., and Allen, G. F. (1969). *J. Chem. Phys.* **51,** 3701.
Lorentzen, H. L. (1953). *Acta Chem. Scand.* **7,** 1336.
Lorentz, H. L. (1965). Statistical Mechanics of Equilibrium and Non-equilibrium, (Ed. J. Meixner), North Holland, Amsterdam.
Marshall, W. and Lovesey, S. W. (1971). Thermal Neutron Scattering. Clarendon Press, Oxford.
Michels, A. and Sengers, J. V. (1962). *Physica* **28,** 1238.
Migdal, A. A. (1969). *Sov. Phys. JETP* **28,** 1036.
Migdal, A. A. (1971). *Sov. Phys. JETP* **32,** 552.
Mistura, L. (1970). *In* Proceedings of the International School of Physics "E. Fermi" on Critical Phenomena, (ed. Green, M. S.) Academic Press, London and New York.
Moldover, M. R. (1969). *Phys. Rev.* **182,** 342.
Moore, M. A. (1970). *Phys. Rev. B* **1,** 2238.
Moore, M. A., Jaswow, D., and Wortis, M. (1969). *Phys. Rev. Lett.* **22,** 940.
Mountain, R. D. (1966a). *Rev. Mod. Phys.* **38,** 205.
Mountain, R. D. (1966b). *J. Res. Nat. Stand.* (M.S.) **70A,** 207.
Mountain, R. D. and Zwanzig, R. (1968). *J. Chem. Phys.* **48,** 1451.
Naldrett, S. N. and Maass. O. (1944). *Can. J. Res.* **18B,** 322.
Noble, I. D. and Bloom, M. (1965). *Phys. Rev. Lett.* **14,** 250.
Onsager, L. (1944). *Phys. Rev.* **65,** 117.
Ornstein, L. S. and Zernike, F. (1914). *Proc. Acad. Sci. Amsterdam,* **17,** 793.
Ornstein, L. S. and Zernike, F. (1916). *Proc. Acad. Sci. Amsterdam,* **19,** 1312, 1321.
Patashinskii, A. Z. and Pokrovskii, V. L. (1966). *Zh. Eksp. Teor. Fiz.* **50,** 439 (*Sov. Phys. JEPT* **23,** 292).
Polyakov, A. M. (1969). *Sov. Phys. JEPT* **28,** 533.
Polyakov, A. M. (1970). *Sov. Phys. JETP* **30,** 151.
Reif, F. (1965). Statistical and Thermal Physics. McGraw-Hill, New York.
Roach, P. R. (1968). *Phys. Rev.* **170,** 213.
Roach, P. R. and Douglas, D. H. Jr. (1967). *Phys. Rev. Lett.* **19,** 287.
Rushbrooke, G. S. (1963). *J. Chem. Phys.* **39,** 842.
Sengers, J. V. (1970). *In* Proceedings of the International School of Physics "E. Fermi" on Critical Phenomena (ed. Green, M. S.) Academic Press, London and New York.
Sette, D. (1969). *Rivista Nuovo Cimento Serie* **I,** *1,* 403.

Sette, D. (1970). *In* Proceedings of the International School of Physics "E. Fermi" on Critical Phenomena (ed. Green, M. S.) Academic Press, London and New York.
Stanley, H. E. (1971). Introduction to Phase Transitions and Critical Phenomena. Oxford, Clarendon Press.
Stell, G. (1968). *Phys. Rev. Lett.* **20,** 533.
Stell, G. (1970). *Phys. Rev. Lett.* **24,** 1443; **24,** 2811.
Swift, J. (1968). *Phys. Rev.* **173,** 257.
Swift, and Kadanoff, L. P. (1968). *Ann. Phys. N.Y.* **50,** 312.
Swinney, H. L. and Cummins, H. Z. (1968). *Phys. Rev.* **171,** 152.
Tartaglia, P., D'Arrigo, G., Mistura, L. and Sette D. (1972). *Phys. Rev. A.* **6,** 1627.
Teaney, D. T. (1966). *In* Critical Phenomena (ed. Green, M. S. and Sengers, J. V.) National Bureau of Stand. (Washington) Publ. 273, p. 51.
Thomas, J. E. and Schmidt, P. W. (1963). *J. Chem. Phys.* **39,** 2506.
Thompson D. R. and Rice O. K. (1964) *J. Am. Chem. Soc.* **86,** 3547
Trappeniers, N. J. and Oosting, P. H. (1966). *Phys. Lett.* **23,** 445.
Trappeniers, N. J., Gerritsma, C. J., and Oasting, P. H. (1965). *Phys. Lett.* **16,** 44.
Tyson, J. A. (1968). *Phys. Rev. Lett.* **21,** 1159.
Tyson, J. A. and Douglass, D. H. Jr. (1966). *Phys. Rev. Lett.* **17,** 472 and **17,** 622.
Uhlenbeck, G. E. (1966). *In* Critical Phenomena (eds. Green, M. S. and Sengers, J. V.) National Bureau of Stand. (Washington D.C.) Miscellaneous Pub. 273.
Van der Waals, J. D. (1873). Thesis, University of Leiden.
Vicentini-Missoni, M. (1970). *In* Proceedings of the International School of Physics "E. Fermi" on Critical Phenomena (ed. Green, M. S.) Academic Press, London and New York.
Vicentini-Missoni, M. (1971). *In* Phase Transitions and Critical Phenomena (ed. Domb, C. and Green, M. S.) Academic Press, London and New York.
Villain. (1968). *J. Physique* **29,** 321, 687.
Weiss, P. (1907). *J. Phys. Paris* **6,** 661.
Widom, B. (1965). *J. Chem. Phys.* **43,** 3892, 3898.
Wilson, K. G. (1972). *Phys. Rev. Lett.* **28,** 584.
Wilson, K. G. and Fisher, M. E. (1972). *Phys. Rev. Lett.* **28,** 240.

Multiplex Spectrometry

JOHN STRONG

Astronomy Research Facility, University of Massachusetts.
Amherst, Massachusetts, U.S.A.

I. Introduction

Recently, applications of multiplexing have brought great advances to spectrometry. The two main techniques are Fourier Transform Spectrometry (FTS) and Hadamard Transform Spectrometry (HTS). Multiplexing is the old art in telegraphy and telephony whereby many messages are transmitted over the same wire simultaneously. Application of multiplexing to spectrometry means a procedure whereby many spectral elements are measured simultaneously by a single detector. This application has its maximum effectiveness with a detector which has the same noise output for a large or small signal: with all the spectral elements in a wide spectral band $(\nu_2 - \nu_1)$ falling on the detector simultaneously, or with only one spectral element, $\Delta\nu$. This is a characteristic of thermal-detectors, such as are used in the infrared spectral region; as contrasted with photon-detectors such as photomultiplier tubes, where the noise is not independent of the signal strength but is proportional to the square root of it.

Dr. Pierre Connes has said that the techniques of presently available Fourier transform spectrometry using two-beam interference with up to 2 m path difference could be extrapolated to any path difference, that the concept of instrument-limited resolving power is thus obsolete and the only limitation left is the true one: energy-limited resolving power. This observation, as has been pointed out by Dr. A. T. Stair, Jr., only applies under laboratory controlled conditions.

Since multiplexing with a single thermal detector allows integrating of the partial detector response—that due to one resolution element of the spectrum—over the entire observing time, the detector noise ascribable to such a particular one of many spectral elements is substantially the same as if it alone were measured, integrating the detector noise over the total observing time. If the number of spectral elements simultaneously measured is N, $(\nu_2 - \nu_1)/\Delta\nu = N$, then interferometric multiplexing gives a noise reduction of the order of $\sqrt{N/2}$ over measuring the same spectrum, $(\nu_2 - \nu_1)$, in the same observing time, T, by the conventional procedure of observing the individual spectral elements sequentially, each for a time T/N.

The two main multiplexing procedures (FTS and HTS) would not be possible without the availability of modern computers and computing ingenuities for executing the formidable computations involved in transforming the detector output data into a conventional spectrum plot.

Here we review important optical matters by which these advances are explained, we describe prototype embodiments of each of the two main procedures, outline the history of the development of multiplexing; discuss some of the technological advances on which these procedures critically depend, such as computers, new detectors, and new optical components (periodic film structures and infrared achromatic lenses, etc.); and finally we report on accomplished applications and future promises. The article by Loewenstein (1971) in "Aspen International Conference on Fourier Spectroscopy, 1970," gives a very readable overview of the mathematics of Fourier transform spectroscopy, with other articles in that publication going into full detail.

Ever since Newton's decomposition of white light, Kirchhoff's spectroscope of 1859 and the photographic dry plate, introduced in 1871, spectroscopy has steadily advanced. Rowland's precision diffraction gratings (1882), Michelson's visibility techniques with his interferometer (1891), and the Fabry-Perot interferometer (1901), all mark subsequent epochs of the development of spectroscopic instrumentation.

Development of spectroscopy for the invisible spectra has been punctuated, for the ultraviolet, by the Elster and Geitel photocell (1890) and Schumann photographic plates (1893); and, for the infrared, by semiconductor detectors and optical developments such as synthetic alkali halide crystals for prisms (Strong, 1930). The German PbS detector† and the subsequent post-war

† When I was at Harvard University during World War 2, I received a German PbS cell to evaluate. It had been sent back from Europe by Eisenhower's army. It looked like our old photronic (copper-oxide) cells: a glass-covered disk with two electrodes in the back. My skepticism quickly changed to admiration when I found it was two orders of magnitude more sensitive (uncooled) than my own vacuum thermocouples.

development of cryo-detectors (Levenstein and Low have been leaders in these developments) have predicated the spectacular success of multiplex spectroscopy.

The purpose of spectrometry, to which multiplexing has responded, is to achieve a curve with the strength of spectral elements plotted as ordinates against their frequency as abscissas. It is desired to have such a curve with as high abscissa resolution and as little uncertainty in ordinates as possible.

In the spectroscope the eye sees one evanescent picture in color of the spectral distribution that such a curve represents. In the spectrograph the photographic plate records it permanently. In a conventional spectrometer a slit is scanned across the distribution (or *vice versa*) to give the successive ordinates of the curve. In multiplex spectrometry the detector response is recorded as a function of something done in the spectrometer. In the FTS the radiation to be analysed is divided into two mutually coherent beams and the something that is done is to increase the path difference between these beams in a succession of incremental steps. Then they are recombined before the detector. In HTS an ensemble of slits, arrayed on a strip, is shifted in steps of one slit-width. The strip lies immediately behind a fixed mask that is located at the exit slit position of an otherwise conventional spectrometer. Radiation through all the open slits falls on the HTS detector.

II. Hadamard Transform Spectrometry

Hadamard transform spectroscopy (HTS), which I shall describe first, is a method of multiplex spectroscopy that may be thought of as modified conventional spectroscopy—the conventional exit slit is replaced by an ensemble of juxtaposed slits, each equal in width to the conventional exit slit it replaces. The ensemble of slits is etched in a metal strip, or ribbon. Only half of the ensemble of slits is effective at any one time, those selected by a superimposed mask, and on the average only half of these again are open, the others are closed or opaque. When two or three open slits are adjacent, the multiplex slit will appear to be a single slit that is twice or thrice as wide as the conventional slit. Here we consider the use of a conventional entrance slit as the entrance window of the spectrometer, with the ensemble of exit slits as the exit window.

Figure 1a is a simple example of a multiplex ensemble, for converting a conventional spectrometer to an HTS.

This array of slits provides for three-fold multiplexing. The ensemble of slits covers the range defined by C. Component slits are arranged after the Hadamard code, described below. M is the mask immediately in front of the ensemble, at the position of the exit slit of a conventional spectrometer. The purpose of the mask is to define which of the slits are in play at any one

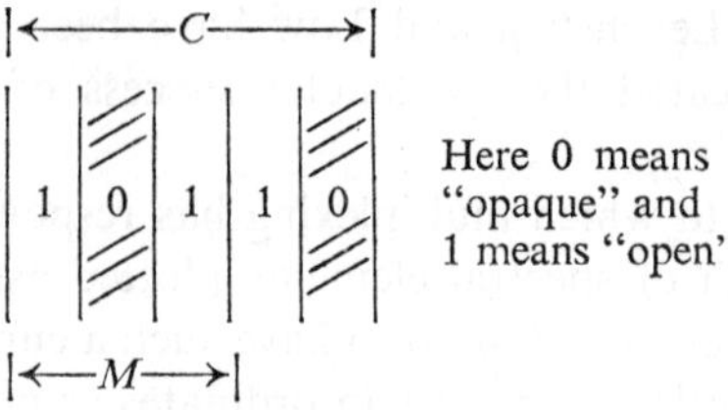

|←—C—→|

| 1 | 0 | 1 | 1 | 0 |

|←M→|

Here 0 means "opaque" and 1 means "open"

FIGURE 1a

$$\begin{vmatrix} 1 & 0 & 1 \\ 0 & 1 & 1 \\ 1 & 1 & 0 \end{vmatrix}$$

FIGURE 1b

$$\begin{pmatrix} 1 & 0 & 1 \\ 0 & 1 & 1 \\ 1 & 1 & 0 \end{pmatrix}^{-1} \begin{pmatrix} d_1 \\ d_2 \\ d_3 \end{pmatrix} = \begin{pmatrix} l \\ m \\ r \end{pmatrix}$$

FIGURE 1c

Multiplex Matrix			Encoding Matrix			Storage Locations l	m	r
1	0	1	+	−	+	$+d_1$	$-d_1$	$+d_1$
0	1	1	−	+	+	$-d_2$	$+d_2$	$+d_2$
1	1	0	+	+	−	$+d_3$	$+d_3$	$-d_3$
					Sum →	$2l$	$2m$	$2r$

FIGURE 1d

time. The mask M is here three times as wide as a single slit; and it exposes exactly three of the multiplex slits. In this instance, one of the slits is opaque.

In operation the energy transmitted by the successive sets of slits that are in play is observed, d. After d_1 is observed, then the ensemble is shifted to the left by one slit width, and the transmitted energy again observed, d_2. After a third shift, yielding d_3, the operation of multiplexing is completed. If the spectral components at the left, middle, and right of the mask are l, m, and r respectively, then

$$d_1 = l + r; \qquad d_2 = m + r; \qquad d_3 = l + m.$$

Here each of the two elements is observed twice, resulting in a signal-to-noise increase of a factor of $\sqrt{2}$. Although this is a very modest advantage, it increases for more complex ensembles.

The basic cyclic array of the ensemble of slits is generated after the first row and the remaining last column from the Hadamard matrix of Fig. 1b where 1 represents an "open" slit, and 0 represents an "opaque" one.

The Hadamard matrix in inverted multiplication solves the combination of the responses d_1, d_2, and d_3 for the spectral components l, m, and r, as is indicated in Fig. 1c.

This equation may be easily solved by computer for a matrix of much higher order if the spectral components are storage registers whose content is determined by an appropriate encoding matrix. Such a matrix for our simple case is shown in Fig. 1d.

The $\sqrt{2}$ signal-to-noise advantage here is modest, but becomes significant with more elaborate, cyclic, multiplex slit arrays.

These more elaborate cyclic ensembles, giving $2^n - 1$ active slits, are derived similarly from square Hadamard matrices of higher order. The successive number of slits embraced by M, for successive matrices, is $N = 3$; 7; 15; 31; etc.

Below, as illustration, we give the first row and remainder of the last column of corresponding matrices to illustrate the total slit ensemble.

$N = 3$ for $n = 2$
101–10
$N = 7$ for $n = 3$
0101110–010111
$N = 15$ for $n = 4$
100010011010111–10001001101011
$N = 31$ for $n = 5$
0000101011101100011111001101001–000010101110110001111100110100

The first digits of the sequences come from the first row of the matrix. The digits after the dash come from the remainder of the last column of the matrix. And the Hadamard slit ensemble is constructed after such information.

We now turn to a sequence that we have used successfully—the sequence derived from an $n = 7$ matrix—yielding a 253 slit ensemble. Here M embraces 127 successive (open and closed) slits, providing for 127-fold multiplexing.

The slit ensemble is shifted 126 times to give the observed data: d_1 to d_{127}. The 127 spectral components passed by the mask measure any particular spectral component only 127/2 consecutive times, because approximately half the slits are opaque. All of the other 127 spectral components that are passed by the mask are measured simultaneously—each one 127/2 times. If the detector is one for which noise is independent of the magnitude of the d's, then one might expect the signal-to-noise gain would be $\sqrt{127/2}$, or an eight-fold gain. Actually, it is $\sqrt{127}/2$ or 5.6 fold.

For a complex spectrum such as a band spectrum, the d's will usually vary some 5% from their mean value. Thus the dynamic range required of the detector system is modest.

III. Fourier Transform Spectrometry

In an FTS, the radiation to be analysed is divided into two beams, a path difference is then introduced between them and they are recombined and focused on a detector. Here they are superposed and they interfere—the detector responding accordingly as component spectral frequencies are in constructive or destructive interference.

Figure 2 shows the two optical devices used for this—the lamellar grating of variable groove depth and the Michelson interferometer with one retro-mirror movable. The first (Fig. 2a and b) divides the radiation to be analysed

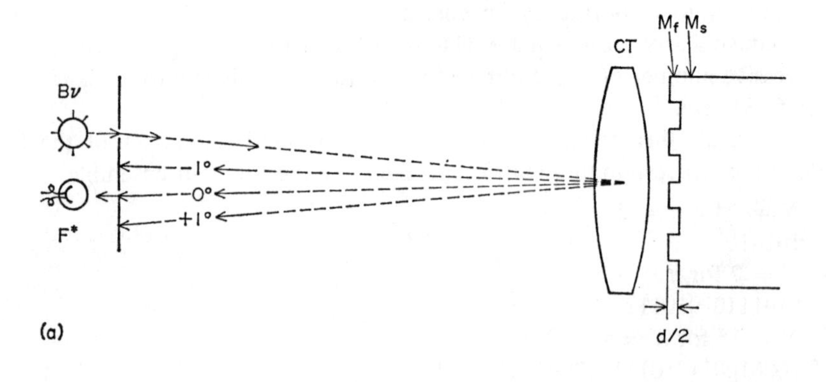

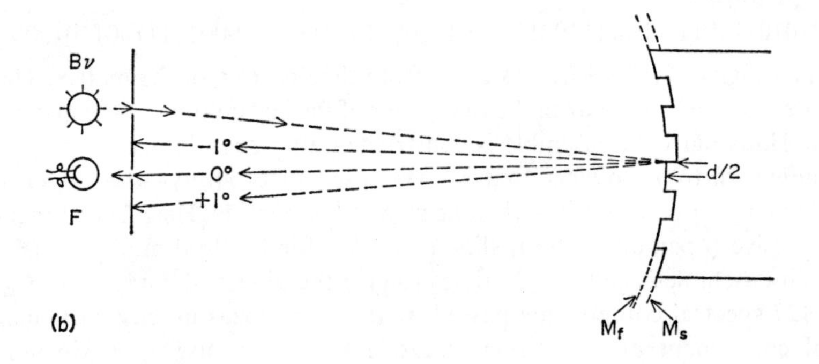

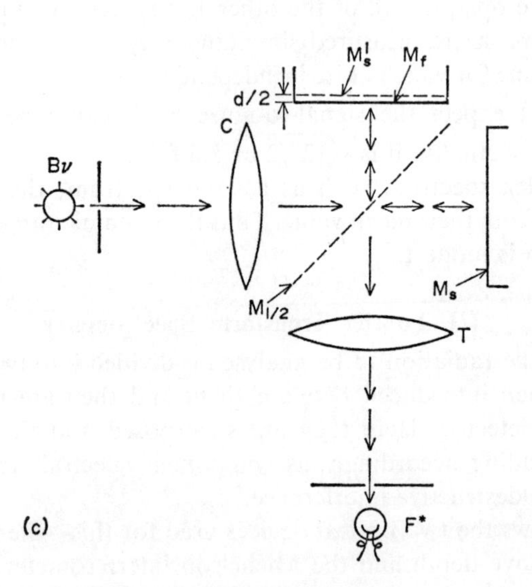

FIGURE 2

by wave-front division and uses the central grating order to activate the detector; the second (Fig. 2c) divides the radiation into two beams by amplitude division and combines them to form two beams of complementary interference—constructive interference for a certain path difference in one of these complementary beams, destructive in the other. $M_{1/2}$, a half-reflecting mirror, serves both as the *beam-splitter* and the *beam-recombiner*.

It should be noted that both devices provide for rejection of the complementary radiation which cannot go to the detector because of destructive interference. When the path difference, d, causes destructive interference for the detector, the complementary radiation, in the case of the lamellar grating, escapes to lateral orders of spectra (formed by the grating on either side of the zero, central order). While in the case of the Michelson interferometer it escapes at a right angle from $M_{1/2}$, returning toward the source.

Both devices use a circular entrance window, e, and a circular exit window, x. We discuss the size of these apertures later, but it is apparent that they are small. When, as in Fig. 2b, the lamellar grating is not flat and the illuminating radiation to be analysed has not been made parallel by a collimator lens or mirror, then several optical niceties are violated, but because the effects of the violations are of secondary importance, it works in practice.

The practice of interferometric spectrometry consists of making an *interferogram*. This is a Fourier transform of the real spectrum made with an analogue device, the interferometer. This interferogram is then transformed into a spectrum, mathematically, by means of a modern digital computer. The detector response to any particular spectral component will be periodic as d is increased. If d is increased at a constant rate, $\dot{d}$, then the frequency of this periodic variation is $f = \nu\dot{d}$. The detector response is the aggregate of the responses for all the frequencies in the analysed spectrum, detected simultaneously.

For either device, as d is increased, the detector response gives only a partial Fourier transform of the analysed spectrum. It is usually displayed as a curve with the detector response plotted as ordinate against corresponding path differences, d, as abscissa. This curve is called the interferogram. If it were continued to $d = \infty$, it would constitute a complete Fourier transform of the spectrum curve. And the mathematically computed Fourier transform of this complete interferogram, to $d = \infty$, would reproduce the original spectrum.

A substantial part of the procedure of FTS is concerned with the consequences of the fact that the interferogram is incomplete. Firstly, it is truncated ($d < \infty$). Secondly, it is not an analytical function, but rather, a series of sample detector responses, one for each step in the motion of the Michelson retro-mirror (or of grating groove depth). And finally, it is sullied with detector noise. This incompleteness, imperfection and intrusion disturbs the validity of representation by the computed spectrum estimate.

Every enterprise is circumscribed by strategic limitations, and the limitation set by the detector and its intrinsic noise has been and remains dominating in FTS. We review, later, how this detector limitation determines energy limited spectroscopic resolving power so that we can understand the roles of multiplexing and interferometry in coping with detector limitations.

IV. Chopping

Pfund (1929) introduced chopping of the radiation to be analysed by an infrared spectrometer. This has been universally adopted because of two advantages that inhere in it. Firstly, it encodes the radiation to be analysed and thus allows its separate detection from other radiation, such as scattered radiation and variable emitted radiation from the detector's surround. Secondly, with chopping, the infrared detector produces an alternating signal which may be easily amplified and then rectified, to return it to a d.c. signal. This is best accomplished, after amplification, by means of a lock-in rectifier. When the variation of the natural emission of the detector surround is imagined as being represented by a Fourier expansion, then only those components within the frequency response band of the detector's readout meter are sensed. In practice such components are almost never significant.

V. Golay Multiplexing

The similar multiplexing schemes developed by Golay (1949) and Girard (1960, 1963), although technically successful, have not been widely practiced. We shall first describe Golay's scheme in a very general way (simply to keep it on the record—it may sometime find an important application).

To understand Golay's scheme, imagine an original pair of entrance and exit slits, passing the frequency ν_0. Next we consider a second pair, the entrance slit being a neighbor to the first, and the second exit slit neighbor to the first exit slit, so, together, the second pair also passes the same frequency ν_0. Golay's scheme consists of an ensemble of such pairs of entrance and exit slits with each slit separately chopped. This chopping produces encoding modulations at the entrance slits and ingenious resolutions at each exit slit such that the a.c. detector system is not responsive to radiations passing through any slit other than the one corresponding to ν_0. All of the other spectral components, other than ν_0, are ignored by the detector because they give modulations outside of its response band. The result: only the frequency ν_0 is detected, and the system passes a large flux at this frequency, proportional to the number of slit pairs, all with spectral resolution unimpaired. To obtain a spectrum, the grating or prism is slowly rotated, or stepped in angle to scan the desired spectral range.

For any spectrometer, a gain, as we have here, in the strength of a spectral element on the detector without sacrifice of spectral resolving power is a *throughput* gain; while the reduction of detector noise by multiplexing, which we do not have here, is called the Fellgett advantage.

VI. Energy-Limited Spectral Power Resolving

The energy limitation is on resolution the dominant one; other limitations, like aberrations, diffraction and size of prisms or gratings, while they may certainly circumscribe ultimate attainable resolving power, are limitations that one can do something about.

In the infrared, inadequate flux of radiation on the detector, or high detector noise, usually limits the minimum setting of slit widths of a prism or grating spectrometer and also the area of entrance and exit windows for a Fabry-Perot spectrometer. These spectrometers are represented stylistically in Fig. 3.

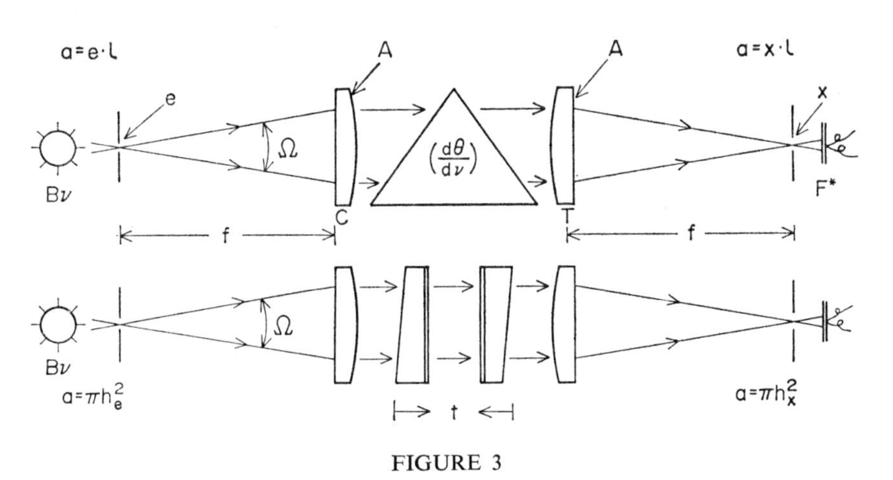

FIGURE 3

The entrance and exit windows of the conventional spectrometers, represented by e and x, are slits of length l. We assume them to lie in focal planes of collimator and telescopic optics, C and T. It serves our explanations to assume that C and T have equal focal lengths. They may be mirrors or lenses. Lying between C and T we consider a dispersing means: either a prism or grating. In either case we ascribe an appropriate angular dispersion, $d\theta/d\nu$. We show only a prism in Fig. 3.

The circular aperture, entrance and exit windows of the Fabry-Perot interferometer are considered, also, to lie at foci, with focal lengths of C

and T that are equal. But here, for the Fabry-Perot, the spectral dispersion is radial, in the focal plane.

A transmission factor, τ, takes account of imperfect spectrometer transmission; it is less than unity because of reflection losses at prism faces and at other optical surfaces, because of body-absorption, or, in the case of the grating, because of the nature of grating blaze efficiency. However, for simplicity, we shall ignore this imperfection and take the transmission of the overall system as unity, $\tau = 1.0$. Furthermore, again for simplicity, we shall suppose the same detector to be used in all cases, a detector that will just give a satisfactory minimum response by some acceptable factor above noise if F^* watts fall on it.

In all cases, if the spectral brightness of the source is B_ν watts $\cdot\,[\Omega \cdot \text{area} \cdot \text{cm}^{-1}]^{-1}$, then a, the minimum entrance and exit window areas that will yield this necessary flux F^*, is implicit in the equation:

$$F^* = B_\nu \, \Delta\nu a \frac{A}{f^2} \text{ watts.}$$

Here the solid angle subtended by the useful area A of C and T, at the windows a, is A/f^2. We now solve F^* for the three cases using the following values for a:

$a_{P.G} = (e \text{ or } x) \cdot l$, where l is the length of e and x;

$a_{F-P} = \pi r^2$, where r is the radius of e and x.

The prism or grating dispersion, $d\theta/d\nu$, gives

$$a_{P.G} = lf(d\theta/d\nu)\,\Delta\nu.$$

In the grating case, if we consider a grating worked in the Littrow arrangement at angle of incidence and diffraction θ, then the grating equation $d\theta/d\nu = \tan\theta/\nu$, which gives

$$a_G = lf \tan\theta(d\nu/\nu).$$

In the F–P case, the radical dispersion is infinite at the center of the Haidinger fringe pattern. But if the apertures are opened to a radius r, the band pass, reckoned from $k/\nu = 2t\cos\theta$, yields $r^2 = 2f^2(\Delta\nu/\nu)$ so that

$$a_{F-P} = 2\pi f^2(\Delta\nu/\nu).$$

Thus, setting the slit widths, or windows a, to yield the energy-limited flux, we may calculate maximum resolutions:

for the prism:

$$\Delta\nu_{P.G} = \sqrt{F^*/B_\nu \, l/f \, A(d\theta/d\nu)\tau}.$$

for the grating worked Littrow, at angle θ:

$$\Delta\nu_G = \sqrt{F^*/B_\nu\, l/f(A \tan \theta/\nu)\tau}.$$

and for the *F–P* case:

$$\Delta\nu_{FP} = \sqrt{F^*/B_\nu(2\pi A/\nu)}.$$

We may now make some cogent comparisons of refraction and diffraction spectrometry with interferometric spectrometry.

Values of $d\theta/d\nu$ for a 60° NaCl prism and for a grating (at $\lambda\lambda \sim 10\,\mu$, in the infrared) are different by a factor of about 10 in favor of the grating. Thus the use of a grating, everything else equal, allows $\sqrt{10}$ or about three times better resolving power than a prism—or it yields an order of magnitude more flux on the detector than a prism at the same resolving power.

Now let us compare the grating with the scanning *F–P* interferometer:

$$\frac{\Delta\nu_G}{\Delta\nu_{F-P}} = \sqrt{\frac{2\pi}{l/f \tan \theta}}.$$

A typical value for l/f is about 1/100 for a grating spectrometer; and $\tan \theta$ is typically about 0.4 for the grating, yielding a ratio of resolving powers of about 40× in favor of the interferometer.

We call the ratio,

$$\frac{F^*}{B_\nu\, \Delta\nu} = a\,\frac{A}{f^2},$$

the effective *throughput* of the system. It is the fact that a can be relatively very large for interferometers (the same relation between a and $\Delta\nu$ holds for the central Haidinger fringe of other interferometers) with automatic encoding by scanning that makes interferometry so effective in coping with the energy limitation problem.

Several details have been neglected which, if accounted, would change our conclusions slightly, but not qualitatively.

Thus FTS using interferometry enjoys both the Fellgett advantage of multiplexing to reduce detector noise and the throughput advantage of interferometry to enhance flux on the detector.

In contrast, HTS does not increase entrance or exit windows for any one spectral element. The gain of HTS comes from reduction of detector noise, arising from the multiplexing. This gain is proportional to the root of N, the ratio of the number of slit widths embraced in the multiplex ensemble to the area of the single exit slit that it replaces.

An HTS is now available commercially (Spectral Imaging Inc., Concord, Massachusetts, USA) which is doubly-encoded—that is, the entrance slit as well as the exit slit is replaced by a Hadamard ensemble. Such a system combines a throughput advantage not unlike the Golay device, with the HTS multiplexing advantage. The total number of data inputs for one spectrum scan becomes $N \cdot n$, where n is the number of slits replacing the entrance slit. Thus the total advantage $\frac{1}{2}\sqrt{N}$ for single encoding is for double encoding, $\frac{1}{4}\sqrt{N \cdot n}$. The factors $\frac{1}{2}$ and $\frac{1}{4}$ come from the circumstance that half the added slits are opaque (whereas all the added area in FTS is open).

The singly encoded advantage is actually

$$\frac{N+1}{2\sqrt{N}}; \quad \text{or} \quad \sim\frac{\sqrt{N}}{2}, \quad \text{if } N \text{ is large.}$$

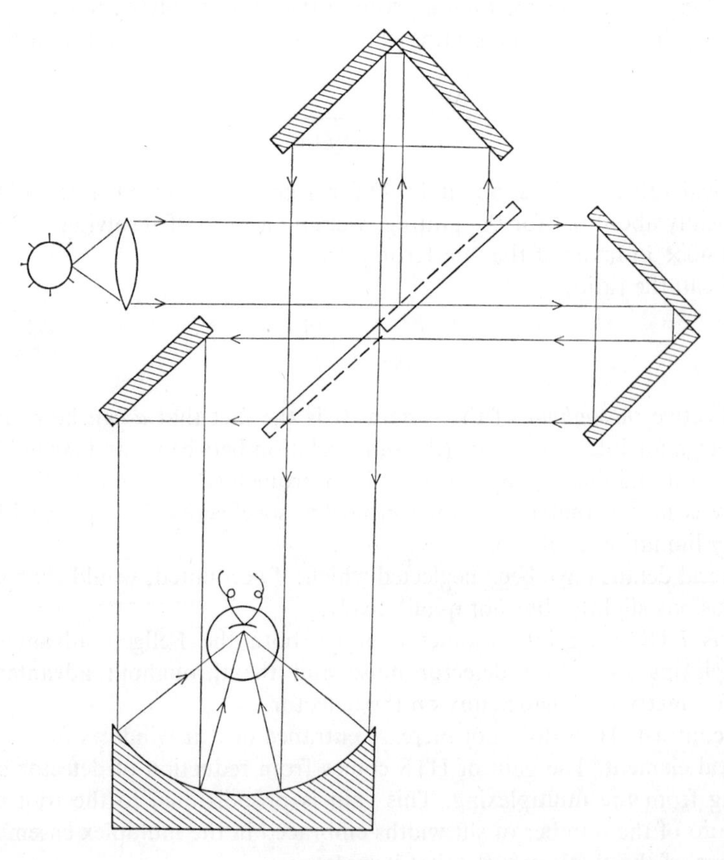

FIGURE 4

In comparison with FTS interferometry, with N spectral elements, the signal-to-noise advantage from multiplexing is $\sqrt{N/2}$, the factor 2 representing the lost complementary beam referred to above. This beam is not lost when using a push-pull scheme (Fellgett's scheme) such as Fig. 4 shows (but becomes $\sqrt{N}/\sqrt{2}$).

In FTS the interferogram available to computation is not an analytical function but merely a sequence of readings of sample detector responses that are determined for a sequence of path differences. Furthermore, the interferogram is incomplete (finite total path difference) and its detector response is sullied with detector noise. Such an imperfect interferogram can be mathematically processed in such a manner as to moderate the consequences of its inadequacies. Apodizing produces an effective scanning function which not only determines resolving power, as does the scanning function of conventional spectrometry, but avoids negative side bands, a characteristic of unapodized processing (which is not found in the scanning functions of conventional spectrometry). Pre-filtering adjusts the computation to the desired spectral range. If the interferometer suffers non-linearities so that the interferogram is not an even function of path difference, then phase corrections are necessary. Even then, the interferometer produces an imperfect Fourier transform of the spectrum and mathematical processing can only moderate defects. If the interferogram were perfect, and not truncated, then its Fourier transform would recover the analysed spectrum exactly.

Figure 5 illustrates some of the imperfections of recovered spectra: several exact spectra are Fourier analysed and then their defective, but still noise-free, interferograms are transformed to recover spectra. By applying various mathematical processing procedures we see the moderation of defects.

We do not need to go into all of the mathematical nuances of recovering the spectrum by Fourier transformation of the FTS interferogram. The reader who wants to know these details can do no better than to refer to the excellent articles in the Aspen Conference report, especially the first by Loewenstein (1971), and also to Vanasse and Sakai (1967).

We will now consider instruments and this will serve as a background to understanding advantages, difficulties, and comparisons of FTS and HTS multiplexing.

The first instrument we describe is our own HTS. It is versatile to the extent that it can be worked either in the HTS mode, or as a conventional double-pass spectrometer.

Figure 6 shows a short portion of the absorption spectrum of water vapor that illustrates the instrument's performance when worked in the double-pass mode without multiplexing. It is adequate to provide good data—data that will give strengths and line widths of water vapor absorption lines (S and γ)

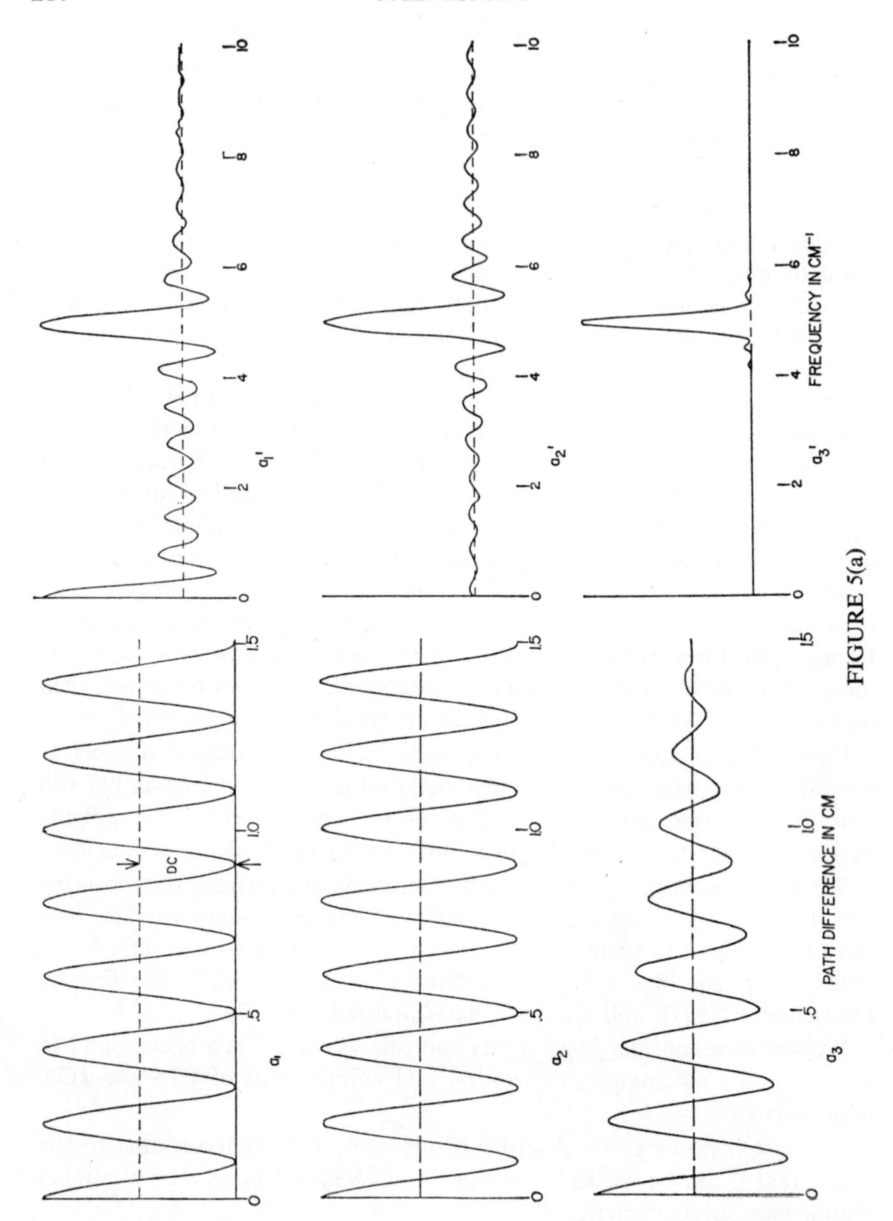

FIGURE 5(a)

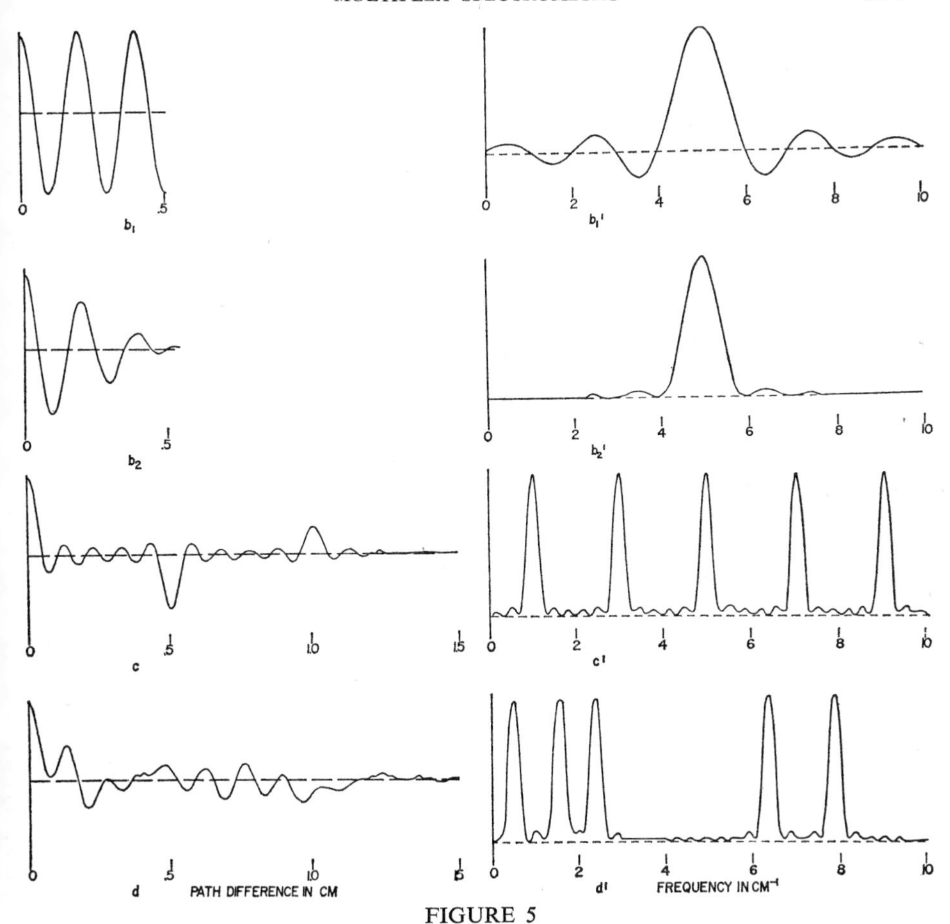

FIGURE 5

Figure 5 shows, on the left, interferograms for various spectra as would be determined by a perfect scanning interferometer, and as modified or not before calculating the transform. The recovered spectra as determined by the calculated Fourier transform of the interferograms are shown on the right.

Figure $5a_1$ is the interferogram of an ideal monochromatic source of 5 cm^{-1} frequency, determined up to a path difference of 1.5 cm. In its calculated transform, a_1', there is a zero frequency component in the recovered spectrum due to the DC component under the dashed line in a_1. Also note the false side lobes and negative intensities of the recovered spectrum. First the interferogram is modified: in a_2 the DC component is eliminated. The result is a recovered spectrum without a zero frequency component, a_2'. Next the interferogram, a_3, has been modified with a triangular apodization function. Note the suppression of the false side lobes in the recovered spectrum, a_3'.

Figure $5b_1$ and b_2 show interferograms of only $\frac{1}{3}$ the path difference of those in a. The recovered spectra b_1' and b_2' are the calculated transforms of b_1 and b_2. Note the threefold greater line widths. b_2 is b_1 apodized.

Figure 5c shows the apodized interferogram of a comb of equally spaced monochromatic lines of equal strength; and c′ shows the calculated recovered spectrum.

Figure 5d is the apodized interferogram of a group of lines of equal strength and unequal spacing; and d′ is the calculated spectrum.

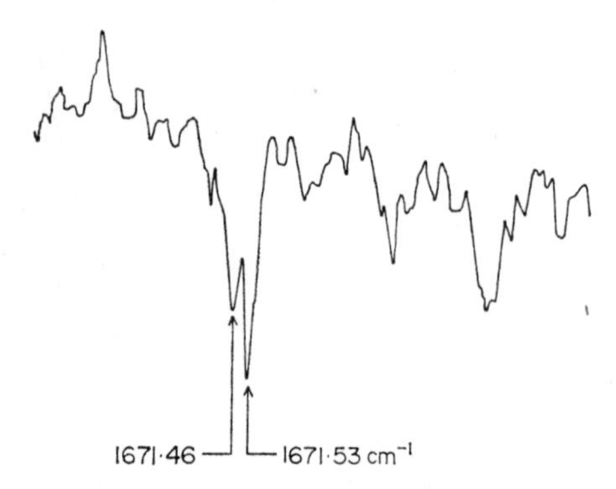

FIGURE 6. High resolution spectra. Water vapor at $P = 0.300$ mm, 92 m path length.

with greater accuracy than they are presently known. We can draw a moral here: One should not use multiplexing unless it is needed. Multiplexing would be needed if one wanted, say, to measure the same lines in emission in the spectrum of the sky. Then, B_ν would be several orders of magnitude less than for the globar source that was used to obtain Fig. 6. For B_ν corresponding to sky temperatures one would need the multiplexing gain of 5.6 of our HTS or the even greater multiplexing and throughput gains that are combined in FTS.

Our HTS uses an echelle grating to get the performance of Fig. 6. As we showed earlier, its dispersion is proportional to the tangent of θ. Here this is large since, for the echelle, $\tan \theta = 2$. The price one has to pay for this advantage (over a conventional infrared grating, $\tan \theta = 0.4$) is that the grating space of the echelle, being some 10 times coarser than for the conventional grating, involves sufferance of serious overlapping of spectral orders. For example, working in the tenth order it is necessary, to be free of overlapping by the 9th and 11th order spectra, to use a narrow band free-spectral-range filter. Such filters must have a uniform high transmission over their transmission bands in order not to waste throughput. Furthermore, a series of such filters is required, with successively contiguous pass bands, to cover a useful spectrum. These filters must be used inside the cryostat in order to block out the strong infrared emission from its outside surround. If the cryostat works at liquid helium temperatures, as ours does, then room temperature infrared emission would greatly over-saturate the detector even within a small cone of exposure. We have devised a compact, entirely nitrogen-cooled prism pre-disperser (Hansen and Strong, 1972; Strong, 1972). This order sorter comes in the optical train immediately in front of the detector. Its cooling includes entrance and exit slits as well as the associated prism, collimating and focusing optics. *LN*-2 cooling is adequate, at $\lambda\lambda \sim 10\,\mu$, to prevent oversaturating the detector in our liquid helium cooled, Low bolometer. The band pass of this pre-disperser is about $\frac{1}{2}\mu$. It is conveniently stepped in synchronism with the echelle (as contrasted with the awkward changing of filters).

The blaze of an echelle is precisely ± 1 grating order wide, and it has the

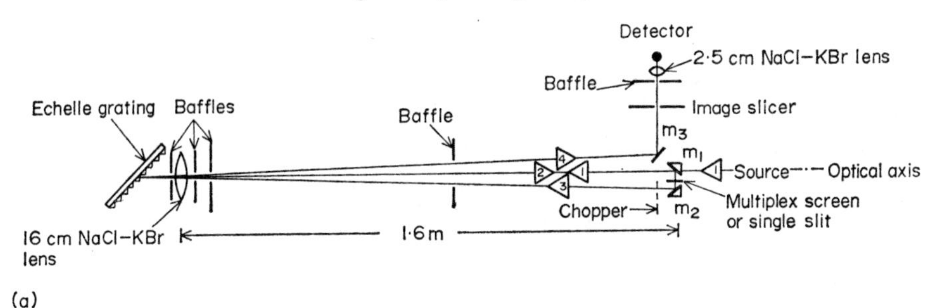

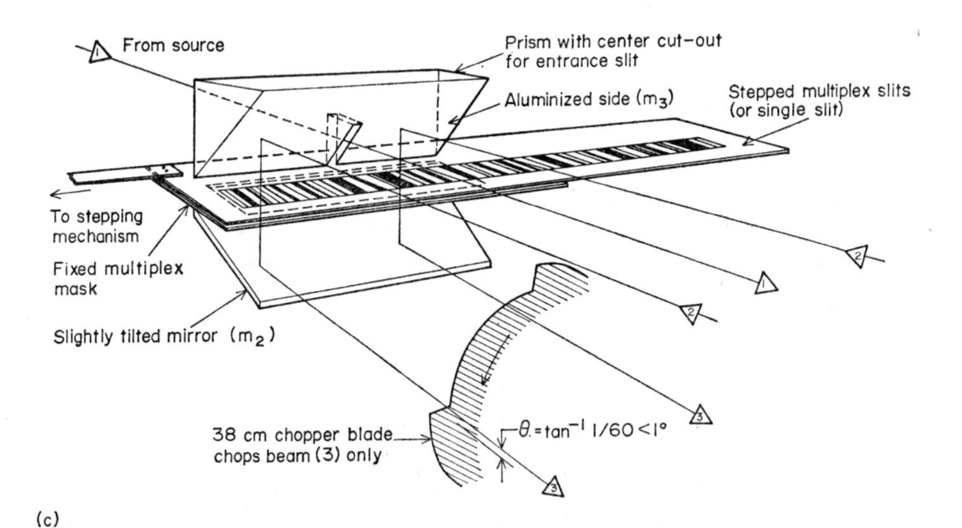

FIGURE 7. (a) Basic HTS spectrometer; (b) equivalent optical path; (c) View of exit focal plane with multiplex slits, corner reflector and chopper.

shape of the diffraction pattern of one echelle groove. Thus $\frac{1}{2}\mu$ is only a fraction of this blaze pattern; and so the blaze efficiency of the echelle is always high.

Figure 7 illustrates our spectrometer (the pre-disperser is not shown).

All infrared detectors should be as small as possible and yet afford a target large enough to receive the spectrometer output. Thus with a wide effective exit window (in our case 50 mm wide), in order to avoid an oversize detector it is necessary to double-pass the echelle: the spectrum passed by the 127 slit exit mask of the HTS is de-dispersed so that it finally images to the same width as the entrance slit (in our case 0.4 mm). However, when the system is used in its conventional mode, we retro-reflect the spectrum to get double, rather than an aggregate zero, dispersion. With double dispersion, $d\theta/d\nu = 2 \tan \theta/\nu = 4/\nu$.

In both instances the chopper is positioned near the middle focus of the double-pass. Chopping there only encodes the pass band that is momentarily being measured, as has been described, so that the detector electronics ignore other flux. This suppresses scattered light—a procedure that has been used extensively in double-passed monochromators (Strong, 1956) that perform, as regards spectral purity, equally with double monochromators.

HTS has several advantages and disadvantages over other procedures of spectrum analysis which we may now enumerate.

Its disadvantage with respect to conventional spectrometry is that a transform of the raw data must be finished before spectrum information is available. This mathematical procedure requires access to a computer, or it requires provision of a special-purpose computer (of some 8000 words memory for a 127 slit Hadamard ensemble). This is not the case when we use our spectrometer in the conventional mode—then spectrum information is currently available.

HTS has the advantage over FTS that the precision required for high resolving power (of 10^{-6} inches) is built into the grating by its manufacturer. Thus machine shop precision (of 10^{-3} inches) is adequate for the slits and their stepping mechanism (our slits are 0.016 in. wide so that ± 0.001 in. is relatively small).

On the other hand, very high precision is required to control the stepping of the path difference in FTS. For high FTS resolving power, each time a step is made it must be made as precisely as if one were ruling a grating groove—there is a subtle analogy between the successive Michelson mirror positions of an FTS scan and successive grating groove positions, considering the grating used near grazing incidence. This precision is $\pm 10^{-6}$ inches.

HTS has a large dynamic range advantage over FTS. Figure 8 shows that the central interference peak, where all spectral components add constructively at the detector, is twice as high as in the wing of the interferogram.

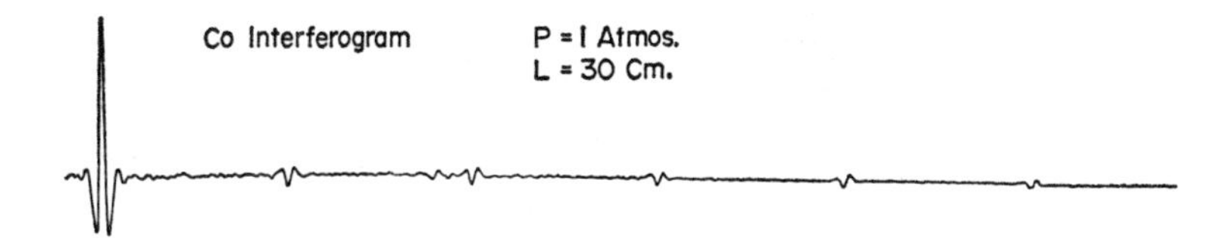

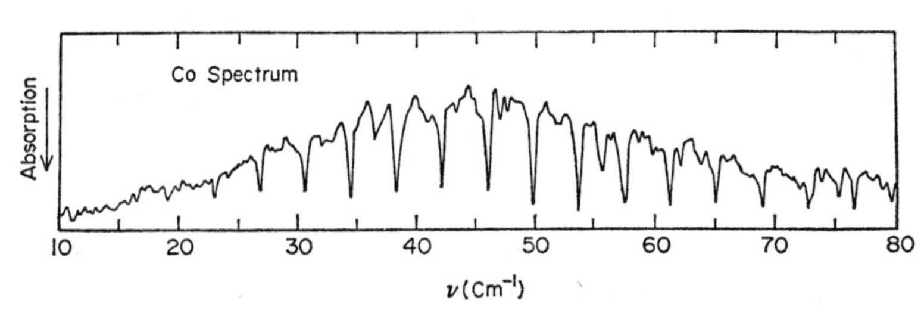

FIGURE 8. (from Loewenstein's Dissertation).

Nevertheless, the small variations in the wing define the strength and width of individual spectral components. Variations of response with path difference due to any one spectral component are the same in the wing as at the center of the interferogram; thus wings carry the same density of information content as at the center, although the wing variations appear not to.

In contrast, the dynamic range required for HTS is much less than for FTS, as we would expect. For example, if we had a spectrum consisting of 50 spectral components, say emission lines of equal strength and ergodic spacing, within the 127 slit span of the pseudo-randomly ordered 64 open slits of the HTS ensemble, then we would expect the energy through the Hadamard matrix to vary by about $\sqrt{64}$, or 8%, as the slits were walked across the spectrum. Actually, when we observe the water absorption band at $\lambda = 6.3\ \mu$, we have a variation of the readouts of about 5%. This 5 or 10% variation lies on top of a total, average detector response that is about the same as response in the wing of an FTS interferogram. The grating is, of course, stepped after each HTS scan, so that an FTS working on the same total spectrum would require a much greater dynamic range. Actually, knowing something of the spectrum, the main part of the HTS responses could be biased away to leave a very agreeable dynamic range. This advantageous use of dynamic range of HTS has long been used in telemetry, especially from space vehicles where the "cost" of telemetering is high.

Another, but minor, advantage of HTS is at least worth mentioning: when measuring a fixed band (the grating not scanned), then an opaque screen may be put over any part of the spectrum which you might wish to ignore. As an example, you could thus ignore an exciting line in a fluorescence experiment.

There is now on the market a doubly encoded HTS spectrometer (Spectral Imaging Inc., Concord, Massachusetts, U.S.A.) that promises to give, with a thermopile detector, the same performance that a corresponding conventional spectrometer could give only with a helium-cooled detector.

Three main types of FTS interferometers are used: (1) those in which the two beams are achieved by amplitude division (e.g. the Michelson interferometer and its derivatives and relatives); (2) those in which the two beams are achieved by wave-front division (e.g. the spherical and plane lamellar grating interferometers; (3) those using vector division (i.e. polarization optics). Mertz (1956) and Sinton (1962) have developed such polarizing interferometers, primarily for the study of stellar spectra, and we do not consider them further here.

It is advantageous to transform oncoming multiplex data as they accumulate. Only the FTS offers this possibility. As this is done, the displayed spectrum improves in resolving power with time as the path difference is growing, or as interferograms for successive path difference scans are co-added, with a running average. Actually, co-adding successive transformed spectra, one for each scan, with an accumulating average is a preferred procedure.

Computing was at first a serious deterrent: to transform for $N = 1024$ spectral elements from an interferogram required $1024 \times 1024 = N^2$ mathematical operations. But, the problem was greatly reduced by the Cooley-Tuckey algorithm. With it, the $N^2 \sim 10^6$ computing operations are reduced to $N \log_2 N \sim 10^4$ operations—a decrease of 100-fold. This decrease, in practice, yields a gain of about a factor 50 in computing time.

Fellgett made several significant beginnings. He combined the multiplexing advantage, using the natural encoding of an interferometer, with the throughput advantage. Golay had pioneered multiplexing to enhance signal strength. At Johns Hopkins University, Rupert (1952) (independently of Professor Jacquinot at CNRS) has recognized the much greater spectral signal strength achievable by interferometric spectrometry. Here high throughput is due to high dispersion.

Interferometry produces two complementary combined beams, as we have already pointed out: one goes to the detector and the other returns to the source, or, one goes into the zero order of the lamellar grating and the other goes into its lateral grating orders. Fellgett reduced his ideas to practice by measuring the reflection of an air wedge between glass plates.

He recorded the reflection of the wedge, as modified by Fizeau fringes, as function of distance from the apex of the wedge. This yielded interferograms.

He used a mercury arc source, and a globar source; and the Fourier transforms of his observations conformed to prism spectra. He did not use the transmission Fizeau fringes, but he pointed out that this was possible, working two detectors in push-pull (see Fig. 4).

The difference between the *interferogram* technique with precise path difference information, and Michelson's *visibility* technique with only approximate path difference information, is that interpretation of a visibility curve is only possible, as was first pointed out by Rayleigh, if the analysed spectrum is a narrow spectral band (such as a single emission line) and also symmetrical about its central frequency. With the interferogram technique there is no such restriction.

In spite of all this we were not motivated to take up interferometry because of the computational barrier, although at the time we were working in far infrared spectroscopy (in the region $\lambda\lambda$ 80 to 1000 μ). Here the precision problem of FTS would have been mitigated by two to three orders of magnitude, due to the longer wavelengths. In the far infrared the high throughput of interferometry is indicated because the energy limitation is acute. A main problem arises owing to the relatively enormous unwanted energy emission of the black-body sources that enters the spectrometer together with the far infrared emission. This unwanted energy is short wavelength energy, at and around the wavelength of maximum black-body emission ($\lambda_m T = 2890$). Various techniques were being used: black paper filters; rough mirror filters; Rubens' shutters; etc. In all, we used seven rejection devices and all working together were still somewhat inadequate. Our motivation finally came with the realization that by using an auxiliary interferometer, to function only as a chopper, we could modulate the different orders of the grating at different frequencies. The unwanted stray radiation is modulated at high frequencies, and thus is separable by means of electrical filtering.

The modulation frequency for scanning at a uniform rate, $\dot{d}$, is $f = \nu\dot{d}$. Madden (1954) and Strong (1954) conducted reduction-to-practice experiments with a lamellar grating, constructed to serve as the far infrared modulator.

This modulator was adequately precise for spectrum analysis because of the greater mechanical tolerances at long wavelengths. Later, Vanasse (1959) constructed a lamellar grating that gave fringes down to $\lambda = 12\ \mu$. It is very difficult (and the difficulty is purely mechanical) to make a lamellar grating operate at wavelengths shorter than that.

It is interesting, and sometimes embarrassing, to note how significant ideas can occur to one, but not acutely enough to be fully appreciated. Figure 9 reproduces a page from my notebook showing that lamellar grating interferometry was conceived 40 years ago. The Boltzmann interferometer referred to is, in effect, only one complete groove of a lamellar grating.

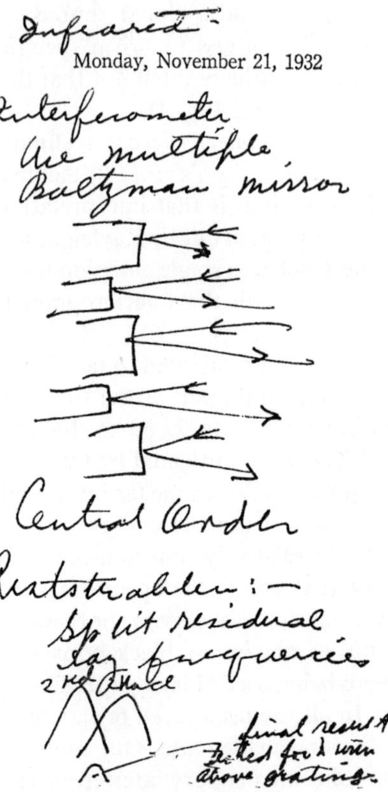

Infrared—

Monday, November 21, 1932

Interferometer
Use multiple
Baltzman mirror

Central Order

Reststrahlen:—
Split residual
ray frequencies

FIGURE 9

Figure 10 shows the lamellar grating as Vanasse adapted it to far infrared interferometric spectrometry. He planned to record the detector output as variable photographic exposures on photographic film from which he could recover the spectrum by scanning this photographic record with a photocell, the output connected to a wave analyser. We were in the midst of this wave-form analysis development when Gebbie joined with us, and he moved us away from analogue to digital analysis. In the early days of this new approach we were indebted to Dr. Finley Wright for all of our computing.

Gebbie also influenced us to invite Dr. Wigglesworth, a mathematician from the United Kingdom, to join our effort. He immediately fulfilled his promise—it was he who first clearly saw and applied apodizing to interferograms, by processing them mathematically with a sync function. (I never knew whether this contribution of Wigglesworth's was original with him, or if he already knew of Rayleigh's precursor work in 1912. I suspect the former.)

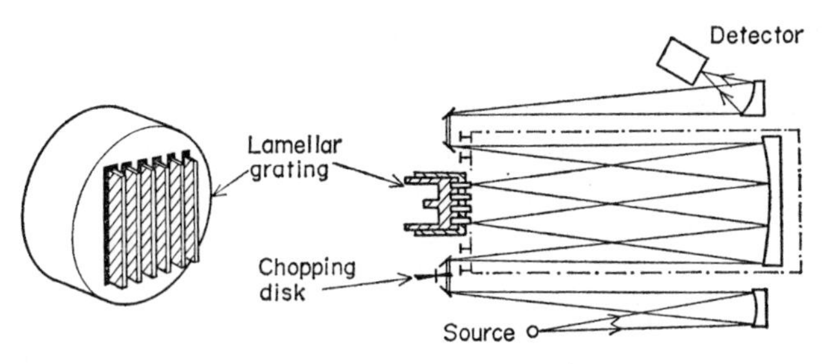

FIGURE 10. Lamellar grating.

After our United Kingdom collaborators left us, and after Loewenstein had joined our effort, we were once accused of visionary indulgences in instrumentation. This accusation was made by an American physicist who is, ironically, now prominent in Fourier spectroscopy.

It is my firm belief that all the early FTS work was just barely indulged by the physics community. But after the Connes (1969) brought out their spectra of the planets, FTS became respectable. I must say, however, that we never experienced anything short of enthusiasm from our original sponsor [embodied in the Office of Naval Research (ONR) person, F. B. Isakson].

In such an historical sketch as this, it would be amiss not to describe the very early work of Rubens and Wood (1911). They had used the high contrast of the index of refraction of quartz in the near and far infrared to isolate a band of far infrared radiation. The contrast is $N \sim 1.5$ for the near infrared and $N \sim 2.2$ for $\lambda\lambda \sim 80\ \mu$. Two crystalline quartz lenses in series isolated this region effectively from the strong short-wavelength infrared. They also determined the wavelengths of the isolated band by observing the interferogram of the transmission of the lenses and scanned a Fabry-Perot interferometer made of two parallel uncoated quartz plates. The index of these plates is so high that the Fabry-Perot interferometer gave approximately two-beam interference:

$$r = \frac{2.1 - 1}{2.1 + 1} = \frac{1.1}{3.1} = 0.355; \qquad \text{yielding } R = 0.126.$$

With this R, the interference maxima and minima are in the proportion of 1 to 0.6. Rubens and Wood did not take the Fourier transform of their interferogram, rather, they calculated Fourier transforms of "guessed at" spectra until they had matched the measured interferogram.

Many parallel scientific and technological advances, some old and some recent, have predicated the very rapid growth of FTS and HTS. I shall outline a few. An appreciation of them must not be lost if these multiplexing procedures are to be promptly brought to the saturation of development that conventional spectroscopy now enjoys.

First and foremost, those who practice spectrometric multiplexing all acknowledge the mathematicians, Fourier and Hadamard; the mathematical physicist, Rayleigh; and the electrical engineers (inspired, perhaps, by the practical needs of an industry) who developed the powerful mathematics of telecommunications. This theory of telecommunications, full-blown, was ready for use in FTS and HTS. FTS and HTS also owe much to the developments of the experimental physicists especially Michelson and Fabry-Perot.

The recent emergence of lasers allows one, using laser light for monitoring, to control the FTS steps by auxiliary interferometry. (This same application of laser light is used currently for control of the stepping in diffraction grating manufacture.) Also, modern developments in ruling diffraction gratings have contributed to FTS procedures.

In FTS the image of the shifted Michelson mirror must remain parallel to the second Michelson mirror, otherwise the central circular Haidinger fringe is destroyed. The first contribution to the problem this tipping poses was made by Peck (1948). He introduced corner cube retroflectors to return the beam parallel to itself regardless of tip. But a lateral shift of the corner cube reflector introduces wave-front shear. The cat's-eye retroflector is a better solution (Connes and Connes, 1966). With it both tipping and lateral shift are innocuous. Use of the cat's-eye retroflector reduces the Michelson mirror shifting problem to a one-dimensional control problem, rather than a three-dimensional one.

In HTS the echelle grating is all important. Gratings were originally (until 1934) ruled on speculum metal. They are now ruled on thermally evaporated films. The first such ruling was made in 1933 on aluminum films that I supplied to Dr. Babcock at Mt. Wilson Observatory. This development was taken up by R. W. Wood at Johns Hopkins and it has become standard for all gratings. The art was expanded to the soft metals, silver and gold, by a special bonding technique using a chromium substrate. This procedure was developed in 1941 by my student J. Winget and myself: the silver vapor source is gradually started when the substrate film of chromium becomes nearly opaque, and as the chromium source is gradually extinguished. This achieves overlapping. By this procedure films of silver can be attached to glass or quartz optical surfaces with such tenacity that a brass rod, soldered to the film, will, when forced, remove chips of the glass rather than pull the film off. Using this procedure Madden (1955) produced a silver film of 13 μ thickness on a large glass rectangular blank (12 by 14 inches) for an infrared grating. This film withstood a coarse deep ruling of 1000 lines per inch. Incidentally, although the silver film was 13 μ in thickness, its surface was free of the bloom sometimes associated with thick evaporated films.

In 1947 Dr. Shane pointed out (personal communication) the merit of crossing two gratings to give a high dispersion spectrum over a long spectral range on a small square photographic plate. We had just ruled two coarse gratings on thick silver films. Both gratings had well defined facets, with corresponding blazes at large and small angles off the grating normal. R. W. Wood and I reduced Shane's idea to practice by crossing these rulings, working one in a high order, to produce high dispersion; and working the other in a low order, to achieve order sorting. Wood (1947) elaborated this idea, using a reflection and transmission grating. The merit of the echelle for HTS lies in the fact that it works efficiently at a high angle of incidence, to produce both high dispersion and high energy spectra; whereas, this high angle is not achievable with efficiency in an ordinary grating.

We made the 254 slots for our HTS ensemble in a steel ribbon of 0.001″ thickness (by photo-etching from both sides of the ribbon). Each slot is 0.25″ long, and the ensemble is a little over 4″ wide. Due to the high dispersion each slot can be 0.016″ wide and subtend only $\Delta\nu = 0.1\ \text{cm}^{-1}$.

For both FTS and HTS, modern detectors are indispensable. Since the first sensitive German PbS cells, the development of semi-conductor detectors has been very rapid. For our HTS system we used a doped germanium chip worked as a bolometer in a cryostat at $T < 2°\text{K}$. It was developed by Frank Low.†

The first cooled, band-pass filter for use as an order-sorter was made by my student Robert Greenler (1955, 1957). He used thermally evaporated tellurium layers for H-films in a HLH-etc. composite, such as Turner had designed. Dr. Heavens, soon after, made similar filters using silicon for the H-films. Now all sorts of film stacks are available commercially for filters and beam splitters.

In our HTS, the wide field (50 mm) is possible because now alkali halide achromat lenses are available for collimator and telescope optics. Mirror optics would not suffice since focal plane fields for mirrors are much smaller than for lenses.

Large alkali halide crystals became available in 1930 for infrared prisms. Until then natural NaCl crystals set the limit of near infrared spectroscopy at $\lambda = 14\ \mu$. With large KBr crystals the infrared region beyond 14 μ was opened up, out to $\lambda\lambda \sim 24\ \mu$.

We did not commercially exploit the process for producing alkali halide crystals, and those who did never realized that they would make excellent achromatic doublets. For example, the *f*/16 lenses for the collimator-telescope optics of our HTS are 15 cm in diameter, stopped to an aperture of 10 cm (*f*/16). Their focal plane fields are 6 cm diameter over which the geometrical blur circle is equal to the diffraction disk or smaller. The proper

† Infrared Laboratories, Inc., Tuscon, Arizona, U.S.A.

combination is NaCl–KBr for the range $\lambda\lambda 5\ \mu$ to $13\ \mu$, and KBr–KI for the range $\lambda\lambda 7\ \mu$ to $24\ \mu$. The specifications of these and other lenses have been described elsewhere (Strong, 1971, 1972). They will be useful in the further development of multiplexing for infrared spectrometry.

All in all many old and new theoretical and experimental developments have converged to attain the present near-mature status of multiplex spectrometry.

Now we turn to some significant applications. The reader who wishes to apply multiplexing to his own problems will certainly want to review them further as they are described more fully in the literature than it is possible in this essay.

The first important new scientific results from FTS came with Connes' (1969) planetary spectra, Hanel's (1972) Mars spectra and Stair's studies of the earth's atmosphere. The planetary emission spectra of the atmosphere of Mars taken in the equatorial and polar regions is the most significant multiplex spectrometry accomplishment to date. Figure 11 shows these spectra.

FTS is useful to study optical constants in the far infrared. Loewenstein

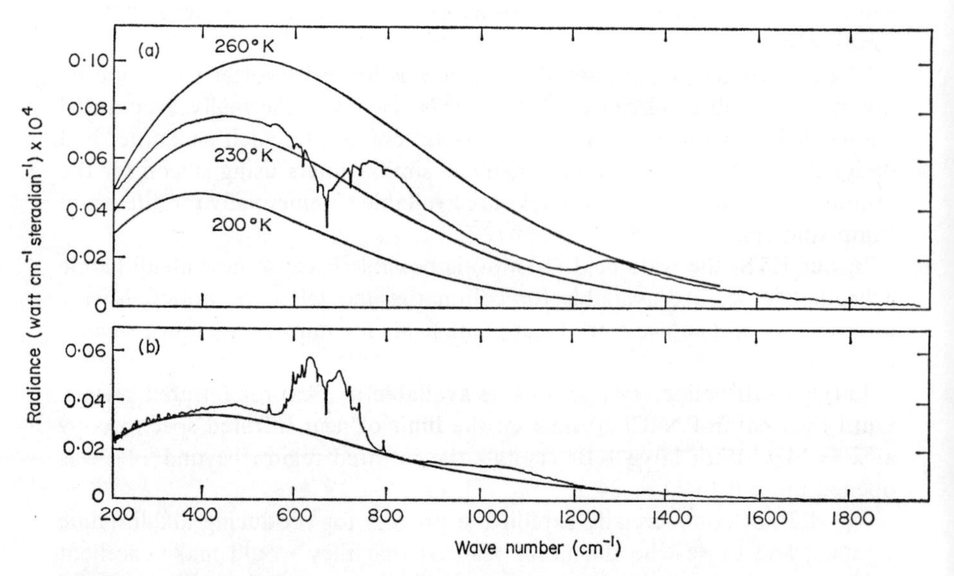

FIGURE 11. (a) Example of nonpolar thermal emission spectra. The spectrum is the average of six spectra obtained from revolution 8 in the region of 18° S, 13° W at about 12.00 local time. Three blackbody curves are included for comparison. (b) Example of polar thermal emission spectra with the south polar cap in the IRIS field of view. This spectrum is the average of six individual spectra obtained from revolutions 29 and 30. The smooth curve is the composite of two blackbody spectra, as described in the text.

(1960) and others (Chamberlain *et al.*, 1963; Bell, 1966) have done this. Such studies are of great importance in solid state physics. FTS is a propitious technique because it can give both amplitude ratios and phase changes for reflection and transmission. Samples are positioned in one arm of a Michelson interferometer (giving asymmetric interferograms). In the past one ordinarily measured the intensity of reflected and transmitted radiation. Now, with FTS, one gets both amplitude and phase. Also, the path difference of a lamellar grating may be immersed in quartz.

There remains much to be done in the field of multiplexing in order to exploit fully its potential and take full advantage of the growing availability of computers adapted to specific tasks. Already conventional spectrometry for analytical chemists is doomed. It remains for the instrument industry to realize this and develop the mature, compact and dependable instruments that will serve this market.

Many ingenuities will no doubt make their way into the art—ingenuities which are anticipated, and some which are not. For example, so-called internal modulation has helped cope with scintillation noise in astronomical applications. This is achieved by an oscillation of path difference of amplitude equal to a half wavelength of the central frequency of the analysed band. The result is in suppression of variations of the d.c. component of the interferogram. This ingenuity reminds us that no one yet has devised an effective double-beaming procedure, a dreamed of device that would make multiplexing immune to variations of source emission in the estimation of the relative strengths of the spectral components.

In 1958 we used a procedure, suggested by W. S. Benedict, that has come to be known as *correlation spectroscopy*. In essence it determined the fraction of a dispersed band spectrum that was transmitted by a special mask as the position of the mask was scanned relative to the spectrum. If, as in our application, the mask consists of slots spaced to match the spacing of the strong absorption lines (or strong absorption features due to groups of lines) of water vapor in its 1.13 μ band, then the transmitted fraction of a spectrum that has been subjected to water vapor absorption will be a minimum when mask and absorption lines correlate. This procedure has also been used with success to estimate trace amounts of the pollutant SO_2 in the atmosphere, working in an ultraviolet SO_2 band. It is required that the gas or vapor to be measured has a unique spectrum for which the correlation is specific, even in the presence of other absorbers.

The extension of this idea—to correlation of interferograms or other multiplex responses, before they are transformed to spectra—is a promising procedure for some particular applications, and the procedure is being developed. Although interferograms have no resemblance to their associated spectra, they are nevertheless as characteristic as spectra.

These are but a few examples of the accomplished successes and future promise of multiplex spectrometry. This essay makes no claims of comprehensiveness (for example, nothing was said of field widening for FTS, etc.). Rather, its purpose has been to set down enough on the how and why of multiplex spectrometry to help the reader to judge whether multiplexing offers any promise for him, to overcome the strategic limitations of his own enterprises.

References

Bell, E. E. (1966). *Infrared Physics* **6**, 57.
Chamberlain, J. E., Gibbs, J. E., and Gebbie, H. A. (1963). *Nature, Lond.* **198,** 874.
Connes, J. and Connes, P. (1966). *J. Opt. Soc. Am.* **56,** 896.
Connes, J., Connes, P., and Maillard, P. (1969). "Atlas des Spectres Infrarouges de Venus, Mars, Jupiter et Saturne." Editions du Centre National de la Recherche Scientifique, Paris.
Girard, A. (1960). *Optica Acta* **7,** 81.
Girard, A. (1963). *Appl. Opt.* **2,** 79.
Golay, M. J. E. (1949). *J. Opt. Soc. Am.* **39,** 437.
Greenler, R. G. (1955). *J. Opt. Soc. Am.* **45,** 788.
Greenler, R. G. (1957). *J. Opt. Soc. Am.* **47,** 130.
Hanel, R. A. *et al.* (1972). *Science, N.Y.* **175,** 305
Hansen, P. and Strong, J. (1972). *Appl. Opt.* **11,** 502.
Loewenstein, E. V. (1960). Interferometric determination of far infrared line widths. Ph.D. Dissertation, Johns Hopkins University.
Loewenstein, E. V. (1966). *Appl. Opt.* **5,** 845.
Loewenstein, E. V. (1971). *In* "Aspen International Conference on Fourier Spectroscopy, 1970" (eds. G. A. Vanasse, A. T. Stair, Jr., and D. J. Baker), pp. 3–17. Air Force Cambridge Research Laboratories; Bedford, Massachusetts.
Madden, R. P. (1954). *J. Opt. Soc. Am.* **44,** 352.
Madden, R. P. (1955). *J. Opt. Soc. Am.* **45,** 408.
Mertz, L. (1956). *J. Opt. Soc. Am.* **46,** 548.
Peck, E. R. (1948). *J. Opt. Soc. Am.* **38,** 1015.
Pfund, A. H. (1929). *Science N.Y.* **69,** 71.
Rubens, H. and Wood, R. W. (1911). *Phil. Mag.* **21,** 249.
Rupert, C. S. (1952). *In* "Abstracts, Symposium on Molecular Structure and Spectroscopy." Ohio State University, Columbus. The appreciation of the multiplex advantage followed elucidation of energy-limited *vs.* diffraction-limited resolving power. Rupert's report was a result of extending these considerations as set forth by Strong, J. (1949). *J. Opt. Soc. Am.* **39,** 320.
Sinton, W. M. (1962). *Appl. Opt.* **1,** 105.
Strong, J. (1954). *J. Opt. Soc. Am.* **44,** 352.
Strong, J. (1956). Optical instrument having through-and-return light path. U.S. Patent No. 2,743,646.
Strong, J. (1971). *Appl. Opt.* **10,** 1439.
Strong, J. (1972). Procedures for infrared spectroscopy. *Appl. Opt.* **11,** 2331.
Vanasse, G. A. (1959). Interferometric spectroscopy in the far infrared (Fig. II-8). Ph.D. Dissertation, Johns Hopkins University.
Vanasse, G. A. and Sakai, H. (1967). *In* "Progress In Optics," (ed. E. Wolf), Vol. 6, pp. 259–330. North-Holland Publishing Co., Amsterdam.
Wood, R. W. (1947). *J. Opt. Soc. Am.* **37,** 733.